Space

SCIENCE, TECHNOLOGY, AND SOCIETY SERIES

(*Formerly Monographs on Science, Technology, and Society*)

2 Edward Pochin *Nuclear radiation: risks and benefits*
3 L. Rotherham *Research and innovation: a record of the Wolfson Technological Projects Scheme 1968–1981;* with a foreword and postscript by Lord Zuckerman
4 John Sheail *Pesticides and nature conservation 1950–1975*
5 Duncan Davies, Diana Bathurst, and Robin Bathurst *The telling image: the changing balance between pictures and words in a technological age*
6 L. E. J. Roberts, P. S. Liss, and P. A. H. Saunders *Power generation and the environment*
7 Roy Gibson *Space*

Space

Roy Gibson

CLARENDON PRESS · OXFORD

1992

Oxford University Press, Walton Street, Oxford OX2 6DP
Oxford New York Toronto
Delhi Bombay Calcutta Madras Karachi
Petaling Jaya Singapore Hong Kong Tokyo
Nairobi Dar es Salaam Cape Town
Melbourne Auckland
and associated companies in
Berlin Ibadan

Oxford is a trade mark of Oxford University Press

Published in the United States
by Oxford University Press, New York

A catalogue record for this book is available from the British Library

Library of Congress Cataloging in Publication Data
Gibson, Roy.
Space / Roy Gibson.
p.   cm. — (Science, technology, and society series; 7)
Includes index.
1. Astronautics.   2. Outer space—Exploration.   I. Title.
II. Series: Science, technology, and society series (Oxford,
England); 7.
TL790.G53   1992   629.4—dc20   91-43590
ISBN 0 19 858343 5

Typeset by Integral Typesetting, Gorleston, Norfolk
Printed in Great Britain by Biddles Ltd, Guildford & King's Lynn

Acknowledgement

Special thanks are due to Norman Longdon of the European Space Agency, for having arranged all the illustrations and for his most helpful comments.

Montpelier   R.G.
March 1992

Contents

Abbreviations and acronyms

ALTP	Advanced Launcher Technology programme
ALS	Advanced Launch System (USA)
AMI	Active Microwave Instrument
Arabsat	Satellite telecommunications consortium of Arab countries (modelled on Intelsat)
ASLV	Augmented Space Launch Vehicle (India)
ATSR	Along Track Scanning Radiometer
AT&T	American Telegraph & Telephone Co.
BNSC	British National Space Centre
CCT	Computer compatible tapes
CDTI	Centre for Development of Industrial Technology, Spain
CNES	Centre National d'Etudes Spatiales (French space agency)
COPUOS	Committee on Peaceful Uses of Outer Space—United Nations
Cospar	Committee on Space Research of ICSU
CSA	Canadian Space Agency
CSG	Centre Spatial Guyanais (European launch site in French Guiana)
CTS	Canadian Technology Satellite programme
CZCS	Coastal Zone Colour Scanner
DARA	Deutsche Agentur für Raumfahrtangelegenheiten (German space agency)
DFVLR	Deutsche Forschungs- und Versuchsanstalt für Luft und Raumfahrt (German aerospace institute)
DLR	New abbreviation for DFVLR
DoD	Department of Defense (USA)
DSCS	Defense Satellite Communications Systems
DSN	Deep Space Network (NASA worldwide tracking network)
ELDO	European Launcher Development Organization
ELV	Expendable Launch Vehicle

ESA	European Space Agency
ESRO	European Space Research Organization
ESTEC	European Space Technology Centre (ESA's technical centre in The Netherlands)
Eumetsat	European Meteorological Satellite Organization
Eureca	European Recoverable Carrier (ESA project)
Eutelsat	European Telecommunications Satellite Organization
EVA	Extra-Vehicular Activity
FFL	Free-Flying Laboratory (part of ESA's Columbus programme)
FM	Frequency Modulation
GEO	Geostationary Orbit
GIS	Geographical Information System
GPS	Global Positioning System
GTO	Geostationary Transfer Orbit
HST	Hubble Space Telescope
Hotol	Horizontal Take-Off and Landing (British Aerospace single-stage to orbit launcher project)
HRV	High-resolution sensor operating in the visual band
IAA	International Academy of Astronautics
IAF	International Astronautical Federation
ICSU	International Council of Scientific Unions
IMCO	International Maritime Consultative Organization
IMO	International Maritime Organization (present name for IMCO)
Inmarsat	International Maritime Satellite Organization
INPE	Institute for Space Research, Brazil
Intelsat	International Telecommunications Satellite Organization
Intersputnik	International Organization of Space Communications (Soviet-Union-based)
ISA	Italian Space Agency
ISAS	Institute of Space and Astronautical Sciences
ISRO	Indian Space Research Organization
ITU	International Telecommunication Union
IUE	International Ultraviolet Explorer
IVA	Intra-vehicular activities
LDEF	Long Duration Exposure Facility (NASA project)
LEO	Low Earth Orbit

Lox	Liquid oxygen
MARECS	Maritime European Communications Satellite
MMU	Man Maneuvring Unit (USA)
MSS	Multi-spectral Scanner
NACA	National Advisory Council for Aeronautics (forerunner of NASA)
NASA	National Aeronautics and Space Administration
NASDA	National Space Development Agency of Japan
NASP	National Aerospace Plane (USA)
NOAA	National Oceanic and Atmospheric Administration (USA)
NSC	Norwegian Space Centre
OMS	Orbital Manoeuvring System
PSLV	Polar Satellite Launch Vehicle (India)
SAR	Synthetic Aperture Radar
SDV	Shuttle-Derived Vehicle (USA)
SETI	Search for Extra-Terrestrial Intelligence
SOHO	Solar and Heliospheric Observatory (ESA project)
SPK	Soviet cosmonaut man-manoeuvring unit
Spot	Système Probatoire pour l'Observation de la Terre (French programme)
SRB	Solid Rocket Booster
SERC	Science and Engineering Research Council (UK)
SSC	Swedish Space Corporation
SSME	Space Shuttle Main Engine
STS	Space Transportation System (NASA)
TM	Thematic Mapper
TWT	Travelling Wave Tube
WARC	World Administrative Radio Conferences
WMO	World Meteorological Organization
WWW	World Weather Watch

1. Introduction

History

The movements and mysteries of heavenly bodies have fascinated people for many thousands of years. Three thousand years before Christ, the Babylonian astronomers are credited with having made observations of the sky, and by 1000 BC the sun clock was in use in Egypt. This line of research was continued by the Greek philosophers, notably Claudius Ptolemy in the second century AD. Indeed, Ptolemy's theories held sway until the Renaissance. It was then the turn of Nicolaus Copernicus (1473–1543), the Polish astronomer, to revolutionize astronomical theories with his conviction that the Earth rotates on its axis and makes an annual orbit around the Sun. Johannes Kepler (1571–1630) refined these new theories, describing the orbits of the planets as elliptical rather than circular (Fig. 1.1). (Modern assertions that Kepler fabricated some of the data in support of his theory should not be allowed to obscure the value of his contribution.)

From that time onward, a succession of scientists have corrected and extended our knowledge of orbital mechanics. One need only mention Galileo Galilei (1564–1642), Isaac Newton (1642–1727) and, in the second half of the nineteenth century, James Clerk Maxwell (1831–1879), as typical contributors.

In parallel with this scientific development, which was concerned more with understanding the movements of the planets than aspiring to visit them, we can trace the gradual appearance of works which we have come to know as 'science fiction'. Sometimes the authors were themselves scientists who had contributed to the development of modern astronomical observations and theories; others were simply imaginative or well-informed writers.

The first, or at least one of the first, works of science fiction appeared in the second century AD, when the Greek poet Lucian of Samosate wrote of a trip to the Moon. His method of propulsion was novel: the ship was lifted into the air by a most violent storm. Kepler himself indulged in science fiction with his novel *Somnium*, which also describes a trip to the Moon.

Cyrano de Bergerac followed in this tradition by writing in 1649 *L'autre monde*, in which he describes a capsule fitted with sails which is transported

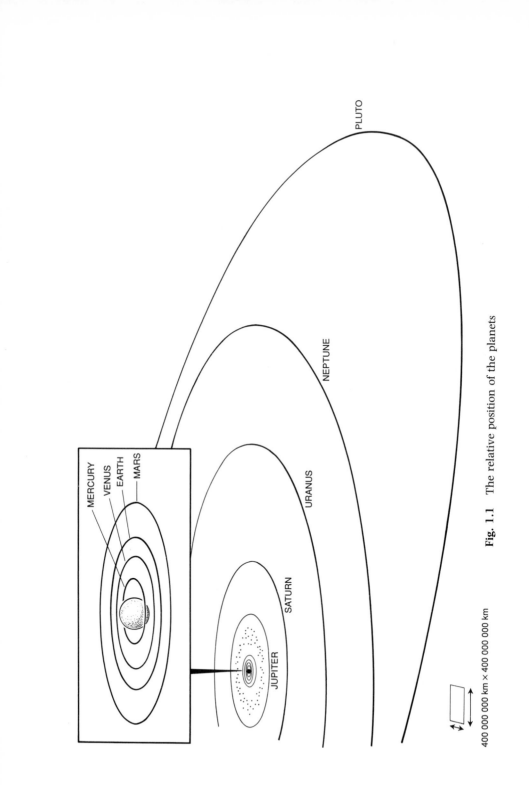

MERCURY
VENUS
EARTH
MARS

JUPITER

SATURN

URANUS

NEPTUNE

PLUTO

400 000 000 km × 400 000 000 km

Fig. 1.1 The relative position of the planets

to the Moon. Unlike Lucian, however, who had remarked meeting three-headed Moon-men, Bergerac disappointingly records that it was not dissimilar to the Earth.

The pace quickened in the nineteenth century with the popular works of Jules Verne and H. G. Wells. Verne's *De la Terre à la Lune* published in 1865, is a not so fanciful account of a trip around the Moon by two Frenchmen and an Englishman, accompanied by two dogs, comfortably installed in a large artillery shell, converted into a space station. Wells's book *The war of the worlds*, which appeared in 1898, describes an invasion of the Earth by Martians, and it too had a great impact on public imagination. Less influential, but nowadays often quoted, Edward Everett Hale in 1869 published the story *The brick moon* in the *Atlantic Monthly*, describing a manned satellite 60 m in diameter which was used to observe the weather and to assist with communications and navigation here on Earth. But probably the most influential of all science fiction writers was a Russian: Konstantin E. Tsiolkovsky (1857–1935), regarded by many as the father of astronautics. A teacher of physics and mathematics in the little village of Kalouga, Tsiolkovsky produced a number of works, which covered most of the main principles of space flight. His classical work was entitled *Rockets into cosmic space*. In 1883, in *Free space* he described a spaceship with an engine using the reaction principle; he appears to be the first writer to have understood—in 1887—that Newton's third law—to every action there is an equal and opposition reaction—is equally applicable in vacuum conditions. By 1912 he was proposing the future use of an electric rocket engine, though he predicted that in the immediate future rockets would have to be powered by liquid propellants, and he specifically refers to liquid hydrogen as a fuel.

Tsiolkovsky's importance lies in his realization that the rocket was the essential key to the exploration of space. The modern development of rocketry, which dates from this period, is dealt with in Chapter 2.

Since the turn of the century, there has been a steady stream of literature, more or less well-informed, on space exploration. Of special interest is a short work (*The world, the flesh and the devil*) by the British scientist, J. D. Bernal, in 1929. The author there wrote about the opening up of the solar system through rocket propulsion. Aware of the ever-present problem of weight limitation, he proposed the use of solar radiation to propel manned spacecraft through the solar system. He was also one of the first to point to the possibility of harnessing solar power for use on Earth, and proposed using the resources to be found in space for the construction and maintenance of manned bases in space.

This tradition of respectable scientists doubling as science fiction writers has continued right up to today. The same Arthur C. Clarke who has written so many works of science fiction and film scripts in the same genre, in 1945 also published an important article on the use of an artificial Earth satellite as a means of relaying radio signals to and from the Earth. But more of this in Chapter 6.

Since, as Tsiolkovsky noted, rockets are essential for the entry into space, it is appropriate at this stage to put progress in other sciences on one side and look specially at the development of rocketry from the days of their appearance on the battlefield as frightening and rather unreliable weapons, up to the construction of the multi-stage monsters now able to put a payload of several tons into orbit.

Although the Chinese are known to have used rockets as long ago as the thirteenth century—if not before—their first significant use, so far as an English author is concerned, was in the defeat of the British army by Tippu, the Sultan of Mysore in the two battles of Srirangapatna in 1792 and 1799. The successful use of this weapon drove Sir William Congreve the younger to design and build rockets at the Woolwich Arsenal. Many tens of thousands of varying sizes, up to 150 kg, were produced and used, for example, from ships in the bombardment of Boulogne in 1806 and even—though not decisively—at the battle of Waterloo in 1815.

The British did not, however, pursue this apparent lead in rocket technology (a trait which was to be repeated in the next century), and the next development occurred in Russia. Nikolai Ivanovich Kibalchich (1853–1881), a contemporary of Tsiolkovsky, wrote in 1881 about a manned rocket platform propelled by gunpowder cartridges. This contribution to human knowledge was, however, dramatically curtailed by his execution as a rebel against the Tsar. White awaiting execution, he continued his theoretical work and wrote about controlling the flight of the rocket by changing the angle of inclination of the motor. Fortunately, there were others to take up the challenge (not only in relation to revolution, but also rocketry).

Yuri V. Kondratyuk (1897–1942) was another important Russian contributor. He was probably unaware of the works of Tsiolkovsky, and he worked out the main equations associated with rocket flight independently. Contemporaneously, there was Friedrich Arturowitch Zander (1887–1933), one of the first engineers to devote his entire professional life to rocket engineering. In the year of his death the first Soviet liquid propellant rocket engine was launched to an altitude of 400 m; it was fuelled by liquid oxygen and alcohol. This was a product of the Gas Dynamics Laboratory established in 1928 in Leningrad. In the next few years, tests were performed on various models of rockets, some of which were used as battlefield missiles in the Second World War. From this stable one outstanding design engineer, Sergei Pavlovich Korolev (1906–1966), emerged to become the father of modern Soviet rocketry.

Developments were proceeding at the same time in Germany and the United States. Hermann Oberth (1894–1989), a Germany physicist, was above all a fervent enthusiast for space. In 1923 he published a book proposing the use of liquid propellants in rockets, and under his aegis the first such German rocket was launched in 1931. Oberth continued the tradition of combining serious scientific study with the popularizing of space and space travel. In 1927 Fritz Lang, the famous German film director, engaged him as his scientific adviser for the film *Frau im Mond* (*The woman in the Moon*).

Oberth attracted the young Wernher von Braun (1912–1977) and other talented engineers, and their work soon came to the attention, and later received the funding, of the German military authorities. This led to the development of the so-called V2* missiles, capable of carrying a 1 tonne bomb over 300 km. The first successful mission was flown on 3 October 1942.

Things developed somewhat differently in the USA. The authorities showed little interest in rocketry and it was left to a university professor, Robert Goddard (1882–1945), to demonstrate their possibilities. Starting in 1910 he addressed the many problems associated with the use of liquid rocket fuels. Largely by his own efforts, he managed by March 1926 to launch the world's first liquid-fuelled rocket. True, it reached an altitude of only 12–13 metres, but it was nevertheless a signal achievement. Goddard persisted in his research and experiments for the remainder of his life, always working with a small team and invariably at loggerheads with the authorities. The American industrialist Daniel Guggenheim supported his work during the 1930s, but Goddard declined all offers to merge his efforts with other groups. Nevertheless, he produced the outstanding number of 214 patents during his lifetime. He was in particular responsible for the concept of the multi-stage rocket, each stage of which fires when the previous stage has been exhausted. Without such a device it would be difficult to take heavy payloads into orbit. As early as 1909 he had understood the advantages to be gained from using liquid oxygen, but his work was largely ignored except by a small group of fellow fanatics.

Thus, by the first decades of the century, there were considerable advances in rocket technology (albeit under differing circumstances) in Russia, Germany, and the USA. It is interesting to note that in all these countries this led to the foundation of space societies. In the Soviet Union this took the form of an interplanetary group in Leningrad, organized by Professor Nikolai A. Rynin (1877–1942), and the publication between 1928 and 1932 of a nine-volume encyclopaedia of space travel, including a volume analysing the ideas of Tsiolkovsky. The German equivalent was Oberth's Verein für Raumschiff-Fahrt (Society for Space Travel), which was created in 1927. The American Interplanetary Society (later to become the American Rocket Society) was founded in 1930 and, though perhaps lacking the same

---

* Vergeltungswaffe Zwei (Vengeance Weapon No. 2).

national credentials, the British Interplanetary Society was also started in 1933.

Without waiting for the subsequent chapters, in which developments in the various fields of activity will be examined, the reader may already feel an urge to dip into some of today's science fiction works, in order to have a glimpse of what is in store. The writers of the past hundred or so years have proved to be astonishingly accurate.

The calculations and experiments of the early pioneers were all aimed at determining the velocity needed for a rocket to overcome the effect of the Earth's gravity. It is important to distinguish between the velocity needed for what became known as a sounding rocket (by analogy with the soundings taken by seamen to investigate the depth of water) and that needed to put an object into orbit. The sounding rocket is, if the manufacturers and other enthusiasts will forgive the simplification, a large firework, which penetrates the atmosphere until its motor is exhausted and the rocket is once again captured by the influence of the Earth's gravity and is pulled back to the ground. To make even that possible it is necessary to impart a velocity of 2000 m/s; to achieve an orbit around the Earth a horizontal velocity of 28 000 m/s is required. Small wonder that space experimenters all over the world cut their teeth on sounding rockets.

A sounding rocket consists essentially of a motor (generally, but not exclusively, made of a solid propellant), the scientific experiment package, and a telemetry system to enable the flight to be tracked from the ground. The experiment package is often recovered by means of a parachute system and since, for safety reasons, the rocket launching site is on the coast, a flotation device is also needed. More sophisticated models include the possibility of transmitting data from the experiments during the flight. Experiments which depend on accurate pointing, necessitate the fitting of star or Sun sensors, which activate small jets to orient the payload correctly. Thus, the major subsystems required in an orbiting satellite are to be found in embryonic form in the sounding rocket, including devices to release the nose cone and expose the experiments. This explains why so many satellite engineers and project managers graduated from sounding rocket programmes.

In the late 1960s and early 1970s, sounding rockets provided virtually the only means for scientists to have their experiments flown in space. Relatively few experimenters were lucky enough to be selected for a place on a satellite launch. Some idea of the extent of the use of sounding rockets can be gathered from the fact that between 1964 and 1972 the ESRO* sounding rocket programme alone notched up 183 launches (Figs 1.2 and 1.3).

Similar, and indeed more extensive, sounding rocket campaigns were

* European Space Research Organization—forerunner of the ESA (European Space Agency).

Fig. 1.2   Last minute adjustments to a sounding rocket experiment—by candlelight

undertaken in the Soviet Union and the USA, as well as in other countries such as Canada, India, Brazil, China, and Japan.

It is difficult to overestimate the importance to the world scientific community of these sounding rocket campaigns. Because the timing of the launch can be selected more or less at will, and because the preparation time is a fraction of that needed for even the most elementary satellite, sounding rockets are particularly suited to observing transient events such as magnetic storms, solar flares, and eclipses. A good deal of the ESRO programme was directed to the study of the aurora borealis in the northern latitudes, using launch sites at Kiruna in Sweden and Andoya in Norway.

Satellite engineers tend to speak disparagingly of these small rockets, but it must be remembered that sounding rockets have accommodated payloads of several hundred kilograms—much larger than the satellites of that some period. Nor is their working life over; on the contrary, the recent shortage

Fig. 1.3 A sounding rocket being prepared for launch in Andoya, Greece, in the 1960's

of launch opportunities has rekindled interest in sounding rockets. The larger rockets are now frequently used for microgravity experiments—miniature furnaces and the like—to take advantage of the short period (5–20 minutes) of near weightlessness that can be achieved during the rocket's flight. The government-owned Swedish Space Corporation plans to launch the MAXUS sub-orbital sounding rocket to heights of around 500 km, giving 6–7 minutes of low gravity conditions.

National and cooperative sounding rocket programmes still flourish in many parts of the world, aimed in particular at studies of the dynamic effects in the atmosphere at altitudes up to 100 km. Study of the auroral ionosphere through sounding rocket experiments is still considered useful to the understanding of fundamental plasma processes.

The advantages of being able not only to take a payload for a ride through the atmosphere with a sounding rocket, but also to give it that extra push which allows it to go into an orbit around the Earth became apparent in parallel with the availability of more powerful, multi-stage rockets developed for military purposes. By the early 1950s it had become obvious that the appearance of the first artificial Earth satellite was very close. In fact, it came on 4 October 1957, in the form of Sputnik I, an 83.6 kg satellite launched by the Soviet Union. It was a simple device: 58 cm in diameter and carrying as its payload just two radio transmitters.

Orbits

Once one starts to talk about satellites, it is inescapable to spend some time describing orbits. Not every satellite traces the same path around the Earth; some orbits are more suitable for specific purposes, some are easier to attain. It is a mini-science in itself, and it needs explanation.

Orbits are either circles or ellipses. Circular orbits are defined by the altitude above the Earth; elliptical orbits by their distances from the Earth at their nearest (perigee) and farthest (apogee) points (Fig. 1.4).

One can choose either an 'equatorial', 'polar', or 'inclined' orbit, depending on the angle of the orbital plane with respect to the plane of the Earth's equator (see Fig. 1.5). Thus a polar orbit has an inclination of 90°. The further away from the Earth the orbit, the less the velocity needed to maintain the satellite in orbit. With a velocity of 8 km/s a spacecraft in circular orbit at an altitude of a few hundred kilometres completes one circuit of the Earth in about 90 minutes. Orbits up to an altitude of around 5000 km are referred to as *Low Earth Orbits* (LEO). As Arthur Clarke demonstrated, there is an altitude at which the orbital period is the same as the Earth's period of rotation: one day. This occurs at an altitude of 36 000 km where, in an equatorial orbit, the satellite appears to be stationary above the Earth. In this orbit, a velocity of only 3 km/s is required. Such an orbit is known as a *Geostationary Orbit* (GEO). Satellites which are in circular orbit at 36 000 km, but not in the equatorial plane, will describe a figure of eight in relation to a fixed point on Earth. This is termed a *Geosynchronous Orbit*.

It should be noted that all satellites (except those in a polar orbit) are launched towards the east, that is to say, in the same direction as the Earth's rotation.

One sometimes hears the terms *semi-synchronous* and *super-synchronous* orbits. The former refers to a circular orbit at an altitude of 20 000 km, where it has an orbital period of 12 hours. The latter is a generic term for all circular orbits at altitudes greater than GEO.

A *Geostationary Transfer Orbit* (GTO) is the elliptical orbit into which satellites are placed before being boosted into the desired GEO. This additional

Apogee

Perigee

Fig. 1.4   Examples of elliptical orbits—known as highly eccentric orbits

velocity, known as 'Delta V', is provided by a motor attached to the satellite.

An interesting variant is the Molniya orbit, named after the class of Soviet telecommunications satellites of the same name. Their orbit is highly elliptical: 40 000 km apogee and 500 km perigee. Because a satellite in this orbit travels more slowly near to its apogee and very rapidly at the perigee, it spends upwards of 11 hours on one side of the Earth. The inclination of the orbit (63°) has been chosen so that the satellite lingers over the northern hemisphere, thus permitting it to provide telecommunications for the Soviet Union.

All the above orbits are illustrated in Fig. 1.4.

There are, of course, many variants to the classic orbits, and many scientific satellites have to be placed in highly elliptical orbits in order to meet the requirements of their experiments. At perigee heights of 100 km,

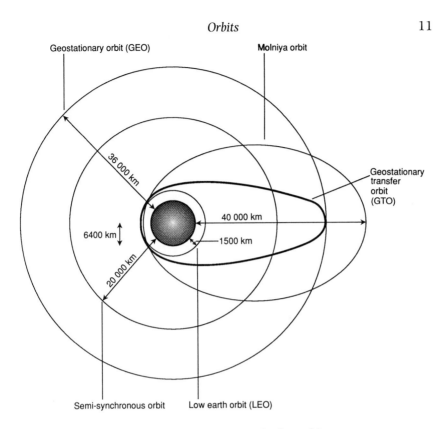

Fig. 1.5   Geostationary and other orbits

the satellite cannot remain in orbit. Final decay usually occurs when the orbital period is reduced to about 87 minutes and the satellite plunges into the lower atmosphere.

As we shall see later on, spacecraft which are intended to visit the planets need a significantly greater velocity. To escape the Earth's gravitational pull, the spacecraft needs to achieve 11.2 km/s, compared to 8 km/s for entering into orbit. If the power of the launcher is insufficient to give this velocity, the last push may be given by an additional motor, which is activated when the spacecraft has been safely parked in a transfer orbit. Clever tricks can also be played by utilizing the gravitational pull of the planets, in order to effect changes of direction which would otherwise be far too prodigal in fuel consumption.

A satellite in orbit around a spherical planet with no atmosphere and far removed from other heavenly bodies would follow its orbit indefinitely without variation. For the planet Earth, however, the theoretical picture is disturbed by three forces:

1. The variation in the Earth's gravitational attraction that results from its imperfect spherical shape. The Earth is measurably flattened at the poles: the equatorial radius is 6378.14 km, whereas the polar radius is only 6356.79 km. There is also a slight asymmetry between the northern and southern hemisphere, which is usually called the pear-shape effect. This, too, causes slight perturbations in a satellite's orbit.

2. Air drag provoked by the passage of the spacecraft through the very thin upper atmosphere, the density of which decreases exponentially with altitude above the Earth's surface. At 350 km, the density is about 10 g/km^3, but with the spacecraft travelling at 8 km/s collisions with the sparse molecules are frequent enough to produce a drag force that is sufficient gradually to decrease the orbit.

3. The forces due to the Sun and Moon, especially their gravitational attraction, but also pressure from solar radiation.

There are numerous other minor perturbations, such as upper atmosphere winds, and all have to be taken into account in calculating satellite orbits and in providing on-board means of correcting the orbit periodically during the lifetime of the satellite. These are known as 'station-keeping' manoeuvres.

An extra word must be said about the geostationary orbit. Because it is especially suited to telecommunication and meteorological satellites, there has been a great demand for geostationary 'orbital slots'. The International Telecommunications Union (ITU), with its 160 member countries, has been obliged to set up a mechanism for allocating positions and for reserving some for those countries which do not yet have their own satellites. The geostationary orbit is thus recognized as an international resource which must be available to all nations. By its very nature, it is a finite resource and its capacity is considerably limited by the need to ensure a minimum distance of 225 km between satellites to avoid interference between their signals. Some nations would wish to see this separation distance increased, not least because satellites are liable to vary their position by $\pm 0.1°$, which at this altitude makes a square of 150 km. There are currently about 100 satellites in the geostationary orbit and perhaps as many again under construction. Apart from the difficulties of obtaining an appropriate orbital position, there is the growing danger of collision between users of the geostationary orbit. Experts now estimate this at around one in a million for each new satellite launched, but some specialists predict that this will change to one in a thousand by the turn of the century.

A more immediate problem, however, is the removal from the geostationary orbit of those satellites whose useful life has finished. Celestial traffic wardens are becoming necessary to ensure that dead satellites neither clutter the geostationary orbit nor gather together in satellite cemeteries, which would be the natural result of gradual degradation of orbit. The European Space

Agency (ESA) set a good example in 1984 by carrying out end-of-life manoeuvres on its scientific satellite GEOS 2 which boosted it into a higher orbit out of the way of the increasing geostationary traffic. Legislation in this field is perhaps too much to expect in the short term, but some form of recommended code of good practice would certainly help.

2. Space transportation systems

Expendable Launch Vehicles (ELVs)

Thus far, rockets have depended almost exclusively on chemical fuel, in either solid or liquid form. Its function is to produce an upward thrust strong enough to take the payload far away from the gravitational force of the Earth, which decreases in inverse proportion to the square of the distance from the Earth. As has already been said, Newton's third law of motion teaches us that the rocket motor's thrust produces a reaction in the opposite, in this case upwards, direction.

In a chemical rocket the reaction between a fuel and an oxidizing agent in a combustion chamber produces hot exhaust gases. The hotter these gases are, the faster they travel and the greater the lifting capability of the rocket. The search has therefore obviously been for techniques of producing the hottest possible reactions, but there are natural physical limits beyond which one cannot go. This in turn imposes a limit on the upward velocity attainable with a simple single-stage rocket—around 3 km/s—far short of the 8 km/s needed to put a satellite into orbit. From this limitation came the concept of multi-stage launch vehicles, each stage igniting when the previous one has exhausted its brief life span. Brief it certainly is: taking the current European workhorse, Ariane 4, the first stage burns for only 20 seconds, the second stage for 124 seconds and the third for 725 seconds. A tremendous burst of power is required to be concentrated into a very short time if the rocket is to escape from the Earth's gravitational pull. Unlike jet engines, which require large amounts of oxygen from the atmosphere, the rocket motor has its own oxidizer and can therefore function in space, better in fact than in normal atmospheric conditions. To give additional power, many modern launchers have additional boosters attached externally to the first stage. These are shown as strap-on solid rocket boosters (SRBs). The capacity of a launcher can be tailored to meet the needs of its payload, by adding or subtracting the optional SRBs. Both the motor stages and the strap-on boosters can be of either solid fuel or liquid propellant. In a solid propellant rocket particles of solid fuel and solid oxidizer are held together with a plastic or asphaltic 'binder'. Once ignited it cannot be extinguished and, although ventilation ports are sometimes provided at the front end of the rocket to dissipate surplus energy, the solid rocket is somewhat inflexible compared to an all-liquid rocket, which in most cases can be turned off and re-ignited so as to achieve precisely the desired thrust for the particular mission.

There are a number of different fuel-oxidizer combinations used in liquid rockets. The more efficient of rockets use liquid propellants with a boiling point far below normal ambient temperatures. These are known as cryogens and require expensive refrigeration for storage. Typical of this class is liquid hydrogen, which boils at a temperature not far above absolute zero (i.e. -273 °C. It is extremely volatile and has a very low density, therefore necessitating bulky tanks aboard the rocket. Other cryogens, such as liquid oxygen (lox) are less volatile and therefore more user-friendly. Kerosene and nitric acid are examples of common propellants of lower volatility which do not need refrigeration. These are known as 'storable' propellants.

Rocket propellant performance is measured by specific impulse: the thrust produced per unit weight of propellant consumed in one second. More explicitly, specific impulse is a shorthand term to describe the number of seconds a particular motor will maintain a thrust of one pound for the consumption of one pound of fuel. Most storable liquid propellants will produce a specific impulse of 250–290 seconds. More efficient motors have reached around 450 seconds, but chemical rockets cannot achieve specific impulses above 500 seconds.

Almost all liquid propellant engines take fuel from one tank and the oxidizer from another. They are therefore known as bi-propellants. It is, however, possible to use a single substance as propellant in which unstable molecules when heated react to form new chemical products and to give off heat. These are known as mono-propellants and, since they require only one set of tanks and pumps, are simpler and cheaper to produce. However, they are less efficient and are mainly used for low-thrust rockets fitted aboard satellites to effect small changes in orientation and spin. Hydrazine, which decomposes into hydrogen and nitrogen, is probably the most common mono-propellant.

For the sake of completeness, a few words should be added about 'hybrid' rockets, that is to say those which contain a solid fuel and binder, but with no oxidizer. A liquid oxidizer, carried separately, is then pumped in through a cavity in the solid fuel when the engine is about to be fired. The oxidizer pump can be stopped at will, and the combustion stops straight away. Hybrid rocket motors are therefore suitable for missions which require the motor to be stopped and restarted several times.

The modern launch vehicle thus consists of a number of stages, each consisting of a rocket motor, tanks, and pumps plus a fairing on the top which provides space for the payloads. It is increasingly common for a launcher to carry more than one satellite, and so it is necessary to design rather sophisticated storage racks under the fairing to accommodate two, and sometimes more, satellites in such a way that they will not suffer damage during the launch phase. In addition, the launcher must accommodate all the 'housekeeping' equipment necessary for the guidance of the launcher in flight, the telemetry which allows the engineers on the ground to monitor the launcher's performance and such devices as the self-destruction package for use in case the launcher deviates from the safety zone agreed for its flight. All this is housed in the equipment bay.

The size and weight of a launch vehicle obviously can vary considerably according to the function for which it was designed. A typical example would be the Ariane 4 launch vehicle (Fig. 2.1) which measures over 50 metres and weighs, without strap-on SRBs, around 250 tonnes. Such a launcher is capable of carrying a payload of only about one or two per cent of its lift-off weight. The precise capacity depends of course on the energy required to reach the desired orbit. The figures for the basic Ariane 4 (known as A40)—typical for this class of launcher—are as follows for the launch of a single satellite:

into GTO*	1900 kg
into Sun-synchronous orbit	2700 kg
into LEO	4600 kg

By adding SRBs, both liquid and solid, six variations of Ariane 4 can be produced, with a proportionate increase in thrust and hence in payload capacity. The most powerful achieves an increased capacity of 120 per cent, i.e. 4200 kg into GTO instead of 1900 kg. Other manufacturers use the same modular technique.

Nowadays, there is no shortage of ELVs. Apart from the range of launchers produced by the old rivals the USA and the USSR, the European Space Agency (ESA) has developed the Ariane family, and both China and Japan produce ELVs of approximately the same capacity. Figure 2.2 shows some representative ELVs presently on the market or in an advanced state of development, and, for comparison purposes, their claimed capacity to put a satellite into GTO.

The reference to 'market' raises the question of price and demand for an ELV. Until comparatively recently, the price was largely a matter between the US launcher manufacturers (Martin Marietta—Titan; General Dynamics—Atlas Centaur; and McDonnell–Douglas—Delta) and Arianespace, the French-based company which markets and launches the Ariane family after each new model has been developed and qualified by the ESA.

---

* Recall that GTO is the transfer orbit from which the satellite is moved into GEO by the use of an additional motor—see p. 9.

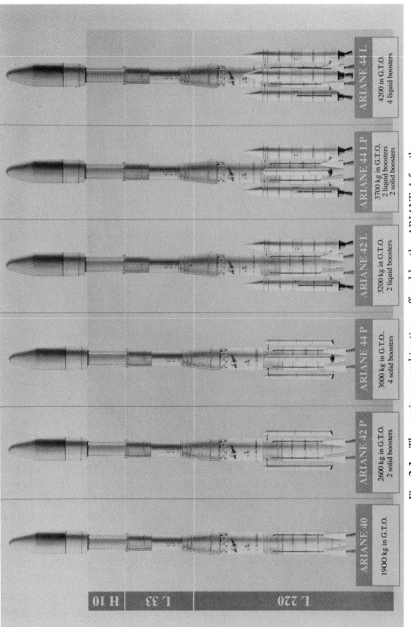

Fig. 2.1 The various combinations offered by the ARIANE 4 family

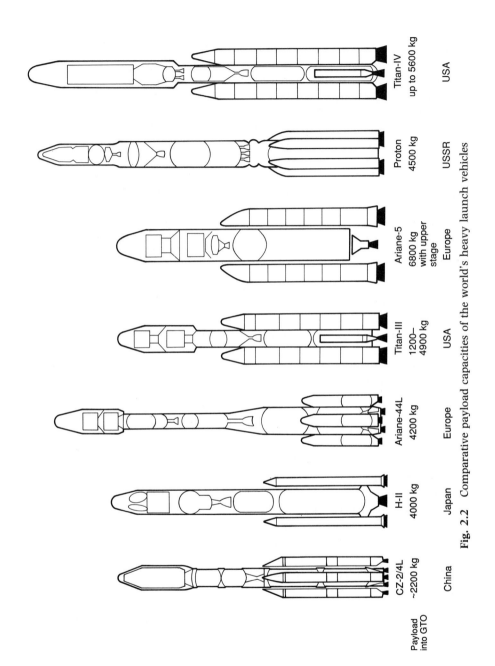

Fig. 2.2 Comparative payload capacities of the world's heavy launch vehicles

Titan-IV	Proton	Ariane-5	Titan-III	Ariane-44L	H-II	CZ-2/4L
up to 5600 kg	4500 kg	6800 kg with upper stage	1200– 4900 kg	4200 kg	4000 kg	~2200 kg
USA	USSR	Europe	USA	Europe	Japan	China

Payload into GTO

This transatlantic rivalry, which led to bitter mutual reproaches over more or less hidden government subsidies, has been disturbed by the offering of Soviet and Chinese rockets for commercial launches. Their launch vehicles had previously been used exclusively for national launches and, given the fundamental differences in their economic systems, there were fears on both sides of the Atlantic that the USSR and China would both be able to undercut the prices demanded by Western ELV manufacturers. The Soviet Union is offering both its Proton and smaller Zenit vehicles, and the Chinese their Long March family. A flow of Western customers to these new launchers has so far been discouraged by the official US refusal to offer any US manufactured sub-systems to be launched on a Soviet or Chinese launcher. The embargo was not quite so explicit in relation to the Chinese ELV, and in spite of a short period of uncertainty after the political repressions in China in mid-1989, US government approval has recently been given for US-built satellites to be launched by Long March for an Australian customer.

Japan is yet another ELV manufacturer due to enter the market as soon as the development of their H2 launcher has been successfully completed. It has recently suffered setbacks, but the Japanese have an enviable record for persistence and consistency in space development, and it is highly probable that their new ELV will be ready for the market within the next three to five years. There are already signs that Japanese industry would like to market this launcher in the same way as Arianespace in Europe. (What the Japanese lack is neither entrepreneurial flair, nor technical competence, but rather a well-placed launch site—on which more will be said later).

India too is investing a lot of effort in developing larger ELVs. Here again, there have been developmental problems, but there is little doubt that their Augmented Space Launch Vehicle (ASLV), will be successfully developed in the next few years. It will, however, have a capacity of only around 150 kg into LEO, and is thus less likely than the Japanese H2 to become a major competitor in the ELV market, at least until the larger Polar Satellite Launch Vehicle (PSLV) is ready, probably near the end of the 1990s.

1989 also saw the first indigenous satellite launch by Israel, and demonstration of a near similar capability by Iraq, although this latter has been destroyed or at least retarded by the Gulf War. But how much does a launch on an ELV cost? With all the possible combinations of launch configurations, satellite weights, different orbits, and the possibility of dual, and even triple, launches, there is no single answer. The short, and true, answer is that the price is still depressingly high, and it is not unusual for a launch to cost as much as the satellite. A representative figure for a launch into GEO would be US$10 000 per kg of payload.

Even at this price, there is still a vigorous demand for ELVs. The commercial market for ELVs now comprises about 20 satellites per year, 80 per cent of which require launch into GEO. Market study predictions indicate that during the period 1995–2000, between 21 and 35 satellites will require launches each year.

Most observers consider that a successful ELV manufacturer needs to average around eight launches a year—say 12 satellites in all. Even taking into account the captive national market—particularly defence launches—it does not seem feasible for all the present and future ELV manufacturers to make an easy living.

The pressure to reduce launch costs has been present since the early space programmes, and many promises have been made that substantial reductions were just around the corner. They have so far not been fulfilled, and the US Space Shuttle is by no means living up to the early claims made for it. For all intents and purposes it can be excluded from consideration in the commercial launching market. It is, in any case, declared US policy to use the Shuttle only for its own defence and NASA programmes, once its existing commercial commitments have been honoured.

The US launcher manufacturers have one considerable advantage over their European rivals: the large number of US government defence launches. Both Delta and Titan benefit from regular defence contracts which are more numerous than the civil launches for which they compete. The corresponding European defence space programme is relatively undeveloped.

Whereas over the past twenty or thirty years the main endeavour was towards ever more powerful launch vehicles, the sheet cost of producing and using them is causing a mild renaissance in the construction of smaller, simpler, and therefore cheaper, launchers and space services.

Both in the USA (with ventures such as American Rocket) and in Europe ('Littleo'), the private sector is developing small ELVs in the hope of creaming off some of the market. In 1991 the Spanish national aerospace establishment, INTA, received government funding for the preliminary design of a launcher capable of putting 50–250 kg payloads into orbit. The concept of the small, almost custom-made launcher is, in theory, attractive. The orbit (and indeed the date of the launch) can be tailored to the requirements of the individual satellite, whereas with dual launches there is inevitably a compromise to be made between the ideal needs of each satellite. Moreover, the cost per kg into orbit should be only a fraction of the cost of the large ELVs.

Such smaller ELVs appear to provide a good solution for payloads of a few hundred kilograms. In practice, however, satellites of this size have mainly been for scientific purposes, and the space science institutes are notoriously short of funds. They depend largely on the various space agencies to provide launches, and are unlikely to be very active buyers on the commercial ELV market. The private sector has so far been more interested in the larger types of telecommunications satellites, and the current trend is for them to weigh near to 2000 kg—far too heavy for the small ELV.

As will be seen later, however, there is undeniably a trend towards the use of very small satellites for environmental and similar experiments. These simple satellites are again a reaction against the growing expense of producing satellites and having them launched. Whether they will be produced in numbers which can support a profitable small ELV industry remains to be seen. An extension of the use of large numbers of small satellites to provide global telecommunications coverage is also being proposed as a commercial venture, and this would of course provide a fillip for the small launcher developers.

Already competition is appearing in the form of an old idea which has recently been refurbished. Why not, the argument used to run, launch our smaller rockets not from the ground, but from an aircraft? In this way, the aircraft pays the initial price of escaping from the Earth's pull, and the rocket has a much easier journey from then onwards. In more precise terms, the rocket needs considerably less fuel to achieve the same altitude than it would have required from a standing start on Earth.

A private US venture, funded by Orbital Sciences Corporation and Hercules Aerospace Co., is now offering to provide such an air-launched rocket, known as Pegasus (Fig. 2.3).

The first operational launch of this 15 m long rocket, which has a gross weight of around 18 000 kg, took place successfully in mid-1990. Pegasus, in its present form will be able to put a total payload of around 350 kg into LEO. Both NASA and the US Defense Advanced Research Projects Agency (DARPA) had small satellites on the first flight and further flights are planned, as well as the development of a more powerful version of Pegasus. The Swedish Space Corporation has also booked a flight for its latest scientific satellite, Freja. A modified B-52 bomber is being used for the first series of flights, and the use of other aircraft is being investigated.

Not the least attractive feature of this airborne launch system is the price: less than 10 million US dollars per launch.

Launching a satellite is not solely a matter of producing enough energy to propel it through the atmosphere. Guidance of the rocket is all-important in ensuring that the satellite is accurately 'injected'; if the inclination, apogee or perigee on the GTO are much different from those planned, the rectification must be made by using the small on-board rockets, thrusters, as they are called. But this uses valuable fuel which is needed to make the regular corrections to the satellite's attitude during its lifetime. An inaccurate injection therefore means a shorter active lifetime for the satellite and, for a commercial satellite, this means a loss of revenue.

ELV manufacturers therefore pride themselves on the accuracy of their launch vehicles. For Ariane 4, for example, Arianespace gives typical standard deviation values of 50 km on the GTO apogee, 1 km on the perigee, and 0.02° on inclination.

More will be said on the subject of accuracy and reliability under insurance

Fig. 2.3 An artist's impression of the Pegasus launcher after release from an aircraft

Fig. 2.4 The launch pads in Kourou, French Guiana for the Ariane family

(Chapter 9), for the majority of satellite owners nowadays seek to insure against the possibility of a launch failure or the injection of their satellite into an unfavourable orbit.

Having the launcher is by no means the end of the story; ELVs require a launch site with expensive installations (Fig. 2.4). The geographical location of this launch site is also of importance; the nearer the equator, the greater the advantage gained from the Earth's rotation. For satellites destined for geostationary orbits, this can result in a significant saving in rocket fuel. For example, a rocket from Kourou in French Guiana (latitude 5°23′N) can carry about 17 per cent more payload than a similar rocket launched eastwards from Cape Canaveral (latitude 28°30′N).

Safety considerations are equally significant determining factors; it is unacceptable for a rocket to have to pass over densely populated areas during its ascent. Some launch sites have more unusual constraints. In Kiruna, Sweden, the authorities have to provide shelters for semi-nomadic Lapps, as well as radios to enable them to receive warnings of impending sounding rocket launches. The Japanese are limited in their use of the Tanegashima launching site to two one-month periods per year, in order not to inconvenience the local fishermen, who have to be indemnified for loss of catch.

This explains the continued search for equatorial sites. Hawaii has for some time been looking into the possibility of building what is now coming to be called a 'space-port', but they have not come so far as the Australians—or more precisely the Queenslanders—with their plan to have a space-port constructed on Cape York. Here a private consortium is well advanced with its plans, and there appears to be an agreement with the Soviet Union for the launching of Soviet Zenit rockets from the site—an arrangement that will bring both logistical and political complications. US approval was recently given to an American company to provide project management support to the Cape York project, and this undoubtedly constitutes another important chink in the US/European commercial launch monopoly.

A launch site's basic features are:

(1) a payload preparation area, where the satellites can be re-checked after their journey from the manufacturers and made ready for flight;

(2) a launch pad, from which the launcher and satellite are despatched, with near-by facilities for assembling the launch vehicle and mating the satellite(s);

(3) a control centre from which the launcher is ignited and monitored in flight; and

(4) the technical and logistic facilities, such as computer installations, propellant storage tanks (or small factories capable of producing them), and a wide range of electronic and mechanical workshops.

Many launch sites have, of course, more than a single launch pad, particularly if they have to accommodate more than one type of launch, and it is characteristic of successful launch sites that they be generously endowed with land, to allow the many installations to be placed at a safe distance from each other (The Guiana Space Centre (CSG) at Kourou, used by the European Space Agency for the Ariane family, for example, occupies around 90 000 hectares).

They are often located in rather remote places (Cape Canaveral is something of an exception), and yet need good communications with the outside world, for both the rocket and the payload need to be flown or shipped in with a minimum of handling.

Tracking the rocket after launch requires radar and telemetry stations, on or near the launch site, and, depending on the launcher's trajectory, several down-range stations to enable the trajectory and the in-flight behaviour to be monitored in real time.

All this is labour-intensive; Kourou, for example, employs at least 1000 people. Moreover, highly skilled personnel are needed, and they have to be compensated for spending long periods of time in locations which, by their nature, are remote from the normal amenities of modern life. All this necessarily adds to the costs, and we can begin to have some sympathy for the extremely high costs per kg which have been quoted above. It is by no means a simple operation!

Flight sequence

In order to understand more clearly what happens after the ignition of the ELV on the launch pad, we may take the flight sequence of Ariane 5, the latest of the ESA launch vehicle family now under development, as an example. Not only is it typical of other large vehicles, but it also provides a bridge to two other important aspects of space transportation: recoverable, as opposed to expendable, launch vehicles, and manned space flight.

As will be seen from Fig. 2.5, Ariane 5 is designed so that the two large solid boosters (designated P230, because they each contain 230 tonnes of granular propellant) can be recovered by parachute, re-furbished and re-flown. Moreover, Ariane 5 is a multi-purpose launcher, and not only can fairings of different dimensions be fitted, so as to accommodate different sizes and shapes of payload, but it can also be adapted to carry the manned space-plane Hermes (Fig. 2.6) into orbit. Hermes, the darling of the French space enthusiasts, has been conceived as the means of transporting astronauts to and from an independent European orbiting laboratory. Whether and in what form it will be included in ESA's long-term programmes will be for ministers to decide. Cost increases during the study phases have not added to its chances.

Ariane 5 consists of a lower and upper composite. The lower composite is the same for all missions and consists of a cryogenic main stage (designated H155 because it contains around 155 tonnes of liquid hydrogen and liquid oxygen) and the two P230 solid boosters.

The HMGO/Vulcain engine which powers the cryogenic main stage is ignited first. If the ensuing check demonstrates that the engine is functioning satisfactorily, the two solid propellant boosters are ignited, and this causes lift-off.

After two minutes, by which time the rocket has reached an altitude of 60 km, the two solid boosters are jettisoned. At a height of 110 km the fairing

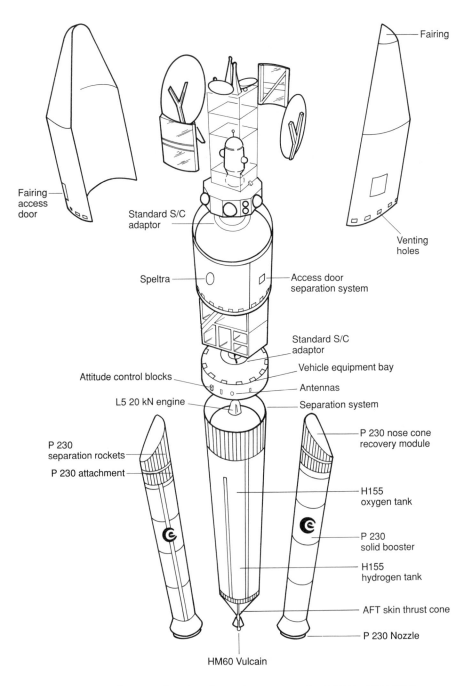

Fairing

Fairing access door

Standard S/C adaptor

Speltra

Access door separation system

Venting holes

Standard S/C adaptor

Vehicle equipment bay

Attitude control blocks

Antennas

L5 20 kN engine

Separation system

P 230 nose cone recovery module

P 230 separation rockets

P 230 attachment

H155 oxygen tank

P 230 solid booster

H155 hydrogen tank

AFT skin thrust cone

P 230 Nozzle

HM60 Vulcain

Fig. 2.5 A diagram showing the main components of ESA's ARIANE 5

Fig. 2.6 The HERMES space plane crossing Europe—an artist's view

covering the payload will be opened and jettisoned—always an emotional time for spectators at a launch. The cryogenic main stage continues to burn for a total of 615 s, which will take the rocket to an altitude of around 145 km. At this point the upper composite separates from the main engine, which gives itself what is termed a de-orbiting boost and re-enters the atmosphere.

The form of the upper composite, which is now left to its own devices, depends on the configuration. In automatic, that is to say unmanned, missions, it consists of a storable propellant stage (L5) for taking the payload into orbit, a vehicle equipment bay containing the brains of the launch vehicle, and the payload covered by a fairing.

The L5 motor is capable of a further 13 minutes' burn, after the satellite is pointed in the right direction, and off it goes, either into its GTO if it is intended for geostationary orbit, or directly into its LEO or Sun-synchronous orbit. The whole process takes about 25 nail-biting minutes.

The Hermes space-plane is intended to replace the whole of the upper composite, and after the separation from the cryogenic main stage Hermes will use its own propulsion unit to move into a circular orbit.

The physical performance limit of chemical propulsions systems having been virtually reached, it is natural that researchers are already looking to other propulsion methods which will be more appropriate to the longer, planetary expeditions which appear on the wish-lists of most space agencies. Using nuclear fission, specific impulses of about 1000s can be achieved, twice that possible with chemical propellants. Nuclear fusion, which takes place naturally in stars such as our Sun, is still beyond our grasp, but a fusion propulsion system would open the way to specific impulses another hundred-fold higher than with nuclear fission.

Apart from the purely technical difficulties of producing nuclear propulsion systems, there are increasingly strong objections from environmentalists. (Nuclear generators have in fact, in spite of such objections, already been used in some Soviet and US spacecraft and indeed in ESA's Ulysses spacecraft, but as a source of power for experiments during a long flight in which solar energy or storage batteries would have been inadequate, and not as a means of rocket propulsion.)

Extremely efficient electrical propulsion systems can also be envisaged, giving specific impulses of around 10 000s, twenty times the chemical propulsion limit, but the snag is that they would require bulky and weighty electrical power supply and conditioning equipment.

Solar and laser thermal propulsion systems have attracted some attention. The energy source is either a large solar collector which focuses sunlight into the combustion chamber, or a powerful remotely operated laser. The specific impulse could by this process certainly be doubled over present limits, but would require large amounts of fuel.

In the USA there are currently a number of interesting research projects directed towards the development of a cheaper way of putting payloads into orbit. One of the most promising is the 'Gas Cannon' project which is being conducted at the Lawrence Livermore Laboratory. This consists of a closed tube, several hundred metres long and about one metre in diameter, in which hydrogen is heated. The payload is intended to sit in the tube until the maximum pressure has been reached, at which point one end of the tube is opened and the payload ejected at what, hopefully, will be velocity sufficient to place it into an elliptical orbit. At this stage of the development, the model payload weighs only 1.5 kg, but the technique has been known for 20 years and extremely high velocities have already been proved possible.

Not, perhaps, of much interest to the satellite project looking around for a suitable launcher in two or three years' time, but certainly a technique worth following.

All rocket propulsion systems entail a trade-off between the weight of the engine and the associated power equipment, and, of course, the weight of the payload which can be carried.

For each specific mission, with its associated energy requirements (Delta V), there is an optimum propulsion system. No doubt the development of new systems will be stimulated if the coming years produce major planetary missions, but for the present, more modest, geostationary missions, we must—at least for the foreseeable future—continue to depend on the now well-understood chemical propulsion systems. Indeed, the demands of the new programmes aiming to establish permanent structures in space (see Chapter 3) have given a renewed impulse for the development of what the Americans often refer to as 'big dumb boosters'—large, simple, chemically-propelled rockets capable of delivering large, unmanned payloads into low Earth orbit. Such a beast exists already in the Soviet Union: the Energiya rocket.

In November 1988, the Soviets launched Energiya, together with an unmanned Buran spaceplane, which closely resembles NASA's Shuttle. Energiya can, of course, be launched in other configurations, and is capable of placing more than 100 000 kg into LEO. It would be able to send a payload of 32 000 kg to the Moon. There is no disputing that this is currently the world's most powerful launch vehicle, and it can be expected to play a major role in the Soviet Union's future space programme—provided that it emerges more or less intact from recent turbulent political events.

The US equivalent, the Saturn rocket which played such an important role in the US manned space programme in the 1960s and 1970s, is no longer available. Both NASA and the Department of Defense have therefore been obliged to solicit bids from industry for the development of new heavy launch vehicles.

The Advanced Launch System (ALS) was intended to have about the same capacity as the Soviet Energiya, but early in 1990 the US Air Force decided not to continue the programme in its present form. Reports are that funding will instead be put into an Advanced Launcher Technology Programme (ALTP) with the aim of producing a new and considerably more powerful launcher later on. It remains to be seen whether the main partner in the ALS will continue its engine development work, or whether it will divert its energy and funding into the development of the so-called Shuttle-Derived Vehicle (SDV): Shuttle C—designed specifically to take 50–60 tons into low Earth orbit. With about twice the capacity of the present Shuttle, such a vehicle would halve the number of flights needed to take into orbit the many different elements required for the US Space Station (see Chapter 3).

We thus find that the spectrum of launchers, from the tiny to the giant size, is still evolving. The number and variety of payloads waiting to be launched is unabated, and the temptation to devise new rockets and more exotic propellants is still proving irresistible to many governments and private investors.

Newcomers, such as Brazil and Israel, are doubtless aware of the military advantages of having an independent lauch capability, and possession of a national launcher still has an undeniable prestige value. Proliferation is inevitable, and we can but wait for the market to determine how many types of rocket are really needed.

Re-usable launch vehicles

The US Space Transportation System—more commonly known as the Space Shuttle—was the first of the re-usable launch systems. Its first test flight took place in April 1981. The Space Shuttle, as will be seen from Fig. 2.7, consists of the aeroplane-shaped Orbiter which, as its name implies, is intended to take the payload into orbit, and also to bring it back to Earth, and the main tank and booster rockets which provide the power to take the Orbiter up.

The Orbiter has a gross lift-off weight of 2000 tonnes, and can accommodate a crew, pilots, and mission specialists numbering seven or eight. Behind the crew station, there is an area for experimentation. This is, however, somewhat restricted, and the major part of the payload is in the cargo bay, which is 18 m long and 5 m in diameter. Of course, this cargo bay is normally only accessible by mission specialists equipped with the cumbersome space-suits, and is used principally for experiments which simply need to be exposed to space by opening the cargo bay doors, or for satellites intended for launch. As will be explained in Chapter 3, ESA provided a pressurized laboratory to fit into the Shuttle cargo bay, and in which mission specialists can work in the comfort of ordinary clothing.

The accident which occurred during the 25th Shuttle flight was followed by public enquiries which often made prime time on the television, and by less spectacular analyses and reconstructions which led to several design changes and a successful re-launch on 29 September 1988. The grounding therefore lasted two and a half years.

The Space Shuttle concept is no doubt a compromise between what NASA

Fig. 2.7 A spectacular launch of the U.S. Shuttle

originally believed to be needed and what Congress would agree to fund. The Space Shuttle is destined to be the US pack-horse for some time to come, and this is not the place to debate its deficiencies, except in so far as this might help to avoid repeating the same mistakes in other programmes.

One important negative consequence of the development of the Shuttle, at least so far as the USA is concerned, was the persistent official discouragement of US ELV development. Pushing the continued development of ELVs was felt to detract from the Shuttle's billing as the multi-purpose space transportation system. Aerospace contractors with large Shuttle contracts did not need to be told twice not to market their ELVs for civil use. The US ELV industry became for, a few years, completely dependent on defence contracts. The Challenger accident in 1986 showed the danger of being totally dependent on a single launch system.

Before the re-assessment following the destruction of Challenger, the Shuttle had been actively marketing for payloads to launch. The prices offered were virtually the same as those for Ariane, but were reckoned to cover less than a third of the actual costs. It is, therefore, not surprising that post-Challenger Shuttle launches are to be reserved for national payloads or for bilateral programmes, such as the manned laboratory Spacelab, produced by the ESA. Once current commitments have been discharged, commercial launches will be left to the ELVs being developed by the US private sector. Successful though it may be in other ways, the Space Shuttle has not provided the economic breakthrough needed to bring down launch costs drastically, as had been hoped and promised.

A number of studies have been undertaken, in various parts of the world, in search of the cheapest feasible means of putting payloads into orbit. In many cases, the concepts go back to the calculations of space pioneers of forty or more years ago, who even then realized the inherent limitations of vertically fired rockets. They, particularly Eugen Sänger (1905–1964), advocated a kind of aerospace plane, capable of taking off and landing like a conventional aircraft, but with a power unit capable of taking it into orbit. The project with which Sänger's name is now linked is in fact a 'two-stage to orbit vehicle': a spaceplane mounted on top of a longer, high performance aeroplane, which can give it, literally, a flying start. The German government, with a contribution from the German aerospace industry, started to fund studies in 1989 at a level (£125m for a period of about three years) which showed their serious intentions. Since then the industries of other European countries have indicated their willingness to take a share of the action.

In the United Kingdom, the Hotol (Horizontal Take-Off and Landing (Fig. 2.8)) project is still being kept alive by British Aerospace, after receiving some small, transient financial encouragement from a government not convinced of the merits of public expenditure on space programmes. More ambitious (some specialists would say 'less realistic') than Sänger, Hotol would be a

Fig. 2.8 A single stage to orbit launcher—HOTOL—designed by British Aerospace

single stage to orbit launch vehicles, using a new design of motor, capable initially of using oxygen from the atmosphere and then switching over to liquid oxygen stored on board. Hotol was designed to be fully automatic, with the possibility of subsequently adding a crew cabin.

The UK government has ceased to give financial support to the Hotol project, but in September 1990 British Aerospace announced a six month study with the USSR aimed at examining the feasibility of launching Hotol from a large Soviet aircraft, Pegasus-style. The results were claimed to be promising, but recent events in the Soviet Union appear to have put a temporary stop to the cooperation.

There is no denying that the success of this project, and all other space-plane projects, depends on a number of technological advances, not to say breakthroughs, but it would certainly have many advantages over current launch systems: the ability to launch from conventional runways, ground support of only a fraction of what is required today, quick assembly of payloads, and turnaround times of days instead of months. Users today would love to have such features, and they seem possible, together with a substantial reduction in price.

There is a good decade of hard work ahead before a firm project emerges, but this prospect has not daunted others from tackling the problem.

In the US, the National Aerospace Plane (NASP) was started in the second half of the 1980s as a joint NASA/Department of Defense (DoD) programme and, although the DoD subsequently withdrew, NASA continues to fight for funding to keep the work going. There is a similar programme being funded in Japan, and the French aerospace company Aerospatiale is also active in the field.

The interest in this type of project is not simply to lower the cost of taking human and cargo payloads into space; there is a realization that the same research and development could lead both to vehicles of military interest and to a successor to the Concorde supersonic airliner. Moreover, the materials research in particular could be of interest to a wide range of industries.

As was the case with the Shuttle, and indeed with many ELVs, designers have found that funding for the development comes much more easily when claims can be made that the result will be a reduction in the launch cost per kilogram. Supporters of the various aerospace planes are naturally not free from such claims. There are, however, a number of reasons why this cost reduction might be more credible than in previous cases. Common to many designs is the use of oxygen from the atmosphere, and this must considerably reduce the weight to be lifted. The Shuttle, for example, must carry all its own oxidizer, and this accounts for about 90 per cent of the weight of the hydrogen and oxygen it carries at lift-off. Another point in favour of the horizontal launcher is the reduced need for facilities on the ground. In the case of the Hotol project, for example, only a few hundred ground personnel would be envisaged, against the army of several thousand needed to launch the Space Shuttle.

But, as they say, nothing is for nothing, and before we can boast of such cost reductions, there is an expensive and long period of research and development to be traversed. The next generation of launchers, in whichever form it materializes, needs new materials capable of withstanding extremely high temperatures, and the development of very sophisticated engines capable of producing speeds of possibly Mach 15* or even Mach 20.

---

* Mach 15: i.e. 15 times the speed of sound, which, at sea level, is 1228 km/h.

3.   Manned space flight

Ever since Gagarin's first flight in 1961, the world has not wanted for critics who demanded to know what good could come from sending people into space. What is perhaps not so well known is that these critics exist equally within the space community. Virtually every nation with a space programme finds itself faced with two factions: the one insisting that more funds be spent on putting people into space and the other arguing that virtually everything can be done with automatic spacecraft (that is the term usually given to unmanned spacecraft) and that in fact astronauts are more trouble than they are worth. Such critics point not only to the considerable advances in robotics and informatics generally, but also the disturbing nature of human presence in an environment where the slightest movement can produce perturbations which significantly degrade the value of the scientific equipment. Spokesmen for the opposing camp—astronauts in the lead—stress the need in space for the unrivalled ingenuity and powers of improvization of human beings.

Nowhere is this dichotomy more obvious than in the USA. Within NASA there have been not only schools of thought, but actual establishments devoted to manned or unmanned space flight. The ensuing rivalry has not always been to NASA's collective benefit. A similar debate rages in the Soviet Union, where Professor Roald Sagde'ev, long-time head of the Soviet scientific space programme, is publicly critical of the tremendous cost of manned space operations.

The reader may be sufficiently unbiased to ask where the truth really lies; there is, however, no easy reply. We can, however, gradually inch our way towards a better appreciation of the problems by establishing a number of ground rules acceptable to all but the unforgiving guerrilla fighters in either camp:

1. Manpower in space is—and is likely to remain—in extremely short supply, and it is therefore prudent to automate functions whenever this can be done at a reasonable cost, and without doing violence to the aspirations of the scientist on the ground.

2. People themselves are valuable experimental material and, as will be

seen in Chapter 5, the results from experiments carried out on astronauts can be of great value.

3. A certain number of operations—and here we are already venturing into disputed territory—require a human presence, and, if the space frontier is to be pushed forward, manned space flight is essential.

All this amounts to saying that there are horses for courses; in other words, there are space programmes where manned presence will clearly give benefits, and these have to be balanced against the undisputed increase in costs—both in manufacture and in subsequent operations. Whether we approve or not, the prestige and media interest attached to manned space flight are also inevitably fed into the equation, so that the outcome is not always so clinically clean as one might wish.

This said, it is time to look briefly at the manned space programmes which have so far been successfully undertaken.

Only the US and the USSR have so far succeeded in sending people into space (though each in its own way has generously provided a space ride for astronauts/cosmonauts of many different countries) and it is fashionable to speak of the 'space race' between these two nations. No doubt the successes and failures of the one had an impact on the efforts and morale of the other, but rather than tell the tale in this competitive fashion, it may be clearer to examine the integrated progress of manned space flight, starting with the early recoverable capsules and moving on to the manned space stations.

In preparation for its manned space programme, the Soviet Union in 1960/61 launched five satellites in the Korabl series; three of these capsules contained dogs. The first manned space flight was on 12 April 1961, when Yuri Gagarin in his Vostok I capsule made a single orbit of the Earth in 108 minutes. This was followed by five more Vostok flights, increasing to 81 orbits, and included the first space flight by a woman—Valentina Tereshkova.

The US followed a similar course, and by February 1962, they too had placed a man, John Glenn, into Earth orbit aboard a Mercury capsule. He finished three orbits, and was then brought down for a sea-landing, whereas the Soviets (no doubt for geographical rather than technical reasons) preferred to recover their capsules on land.

Capsules thereafter started to grow, and by 1964 the Soviet Union's new series of Voskhod capsules could squeeze in three cosmonauts, and the rather more comfortable US Gemini series accommodated two astronauts, permitting flights of somewhat longer duration. Frank Borman and James Lovell, for example, in 1965, stayed aloft for 13 days, making 206 orbits around the Earth.

By 1965, too, both the Soviets and the US had started to experiment with sending their astronauts from the relative comfort of their capsules to perform tasks in space. This extravehicular activity (EVA), as it is prosaically termed, was later put to good use in effecting running repairs to spacecraft and—much later—to actually manhandling large satellites into a cargo bay for bringing back to Earth.

Around this time, there was a divergence in the directions of the two giant space programmes: the US put its resources into the Apollo programme, dedicated to putting a man on the Moon, whereas the Soviets developed the Soyuz series, intended to lead to the Salyut space station. Both achieved their ambitious objectives in about the same time-frame. The US mission (Apollo 11) which first put men on the Moon was launched on 16 July 1969, and between then and Apollo 17 in December 1972 there were further astronaut landings—even a Moon motor ride—and altogether about 400 kg of material were brought back from the Moon. Meanwhile, the Soviet cosmonauts continued with the Soyuz flights and in 1969 achieved the first link-up between two manned spacecraft.

Two years later, in April 1971, the first space station, Salyut, was launched, and in June Soyuz 11 successfully docked with it, and the crew of three spent 24 days aboard the space station. Tragically, a faulty valve apparently evacuated the air from the Soyuz capsule before it re-entered the atmosphere on the return journey, and all three cosmonauts died of suffocation. This accident introduced delay into the Soviet programme, and it was not until September 1983 that they were able to launch their next manned capsule, Soyuz 12.

The US programme at this stage had better luck, and the period of the Soviet grounding coincides with a very fruitful US improvization: the Skylab programme. Critics of the US Apollo programme reproached both NASA and Congress for not having provided a logical follow-on programme, but it is understandable that the effort and the concentration needed to put men on the Moon precluded much thought being given to future space programmes. Effort in this direction was concentrated on using surplus hardware either from the Apollo programme or from the giant launcher Saturn. This crystallized into a programme to convert the third stage of the Saturn 1B into a manned laboratory, to be known as Skylab. It provided 316 cubic metres of accommodation and working space—more commodious than anything else available for more than a decade. Skylab, weighing 82.5 tons, was launched in May 1973, and on three occasions that year 22 ton Apollo capsules were used to bring up astronauts for periods of 28, 59, and finally 84 days. During these visits, the three-man crews carried out an extremely wide variety of scientific and technological experiments—over 200 in all (Fig. 3.1). The data from these missions filled over 80 km of magnetic tape, and the astronauts managed to take 40 000 photographs of the Earth's surface, and more than four times that number of events on the surface of the Sun—a rich harvest. But of necessity, they had to turn their hands to a number of essential repairs on the outside of Skylab.

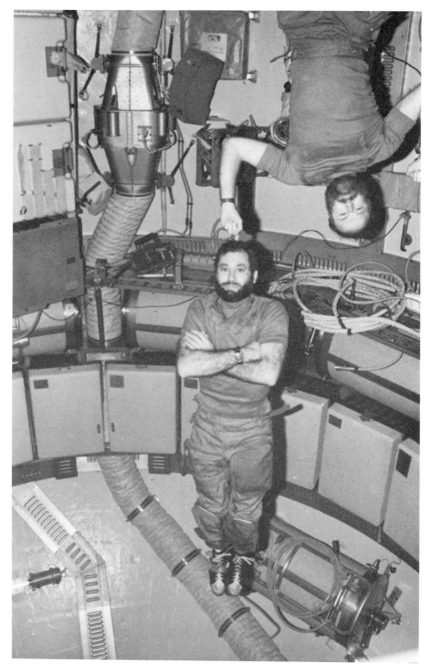

Fig. 3.1 Astronauts Pogue and Carr performing the traditional circus stunts in the weightlessness of SKYLAB

The Skylab programme used three of the four Saturn 1B rockets remaining form the Apollo Moon programme. The last one was reserved for an international space spectacular: the Apollo–Soyuz rendezvous mission in mid-1975. Interesting demonstration of peaceful cooperation though it undoubtedly was, the mission did not add very much to the sum of human knowledge. When one reflects, however, on the difficulties then prevailing in any technical exchanges with the Soviets, it is a wonder that the programme ever came to fruition.

In the remainder of the 1970s, the space headlines belonged to the Soviet Union as they continued to build up and improve the Salyut orbital complex. This developed, in addition to the Salyut station, which was itself progressively upgraded, through the use of three further elements:

(1) Soyuz T transport spacecraft used for the delivery and return of cosmonauts and the transport of cargo;

(2) Cosmos 929-type module used to increase the volume of the station; and

(3) Progress automatic transport spacecraft used for delivering equipment and consumables.

Multiple flights of Soyuz spacecraft were undertaken to take crews, including foreign cosmonauts, to visit the Salyut stations. It is interesting to note that even Salyut 7, the last of the series, had a mass of only 19 tonnes and volume of 90 cubic metres—less than a third of the size of the Skylab space stations which, incidentally, came down through the atmosphere in July 1979.

By 1970, NASA had started the long process to win approval for its new programmes. Originally, the planners had hoped to have agreement for both a recoverable Space Shuttle and an associated space station, but only the Shuttle survived the approval process.

For a Shuttle launch, the three Space Shuttle Main Engines (SSMEs) are first ignited, and if no anomaly appears, the SRBs are ignited shortly afterwards; otherwise the SSMEs can be shut down. The SRBs, once ignited, cannot of course be shut down, and a few minutes after launch they are ejected and returned to Earth by parachute for refurbishing and re-use. After about eight minutes of flight, the SSMEs shut down and the external tank separates from the Orbiter, which continues into orbit, powered by the Orbit Manoeuvring System (OMS). Unlike the solid boosters, the external tank is not recovered. Once in its orbit, about 200 km above the Earth, the Orbiter can commence its work: launching, or even recovering, satellites, carrying out scientific experiments, or a combination of all these. On the completion of its mission—seven to ten days—the Orbiter fires its OMS engines to leave its orbit, re-enters the atmosphere and glides to a landing on an extended runway (Fig. 3.2).

Work began on the Space Transportation System (STS) in 1972, and the

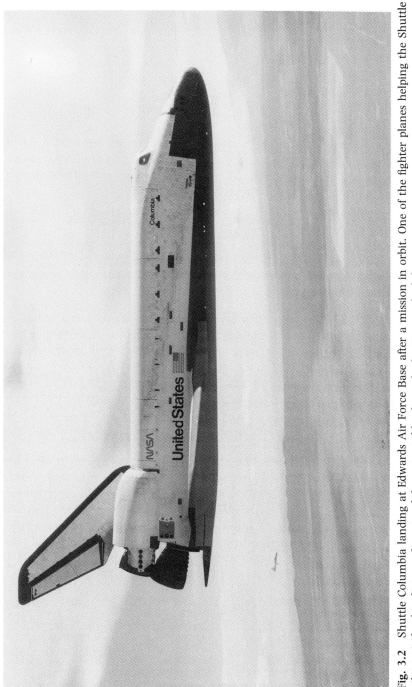

Fig. 3.2 Shuttle Columbia landing at Edwards Air Force Base after a mission in orbit. One of the fighter planes helping the Shuttle pilot to judge his distance from touchdown is visible above the horizon at the left

first launch was in April 1981. NASA originally ordered three orbiters, but—as the whole world surely knows—the orbiter Challenger was lost with its crew in a tragic accident in January 1986. A replacement orbiter has subsequently been ordered.

The US invited European participation in the STS programme, and there were long years of discussions and negotiations, both at governmental level and between the two space agencies—NASA and ESA—before agreement was reached that ESA should develop the Spacelab, a pressurized laboratory designed to fit into the orbiter's payload bay. ESA successfully completed this programme and spacelab flew for the first time on Shuttle flight Number 9 in November 1983. Figure 3.3 shows Spacelab in the final stages of construction in Europe, prior to being shipped to the USA.

Fig. 3.3 ESA's SPACELAB being readied for shipment to NASA for its flight in the cargo bay of the Shuttle

Even before the Challenger accident in 1986, the Shuttle programme was the target for considerable criticism in the US. Opponents strongly disputed that this was the correct way to reduce launch costs, and the rising development costs (not unexpected when one considers the enormity of the problems in the SSMEs alone) fuelled the controversy. A criticism that was perhaps on surer ground was directed against NASA's decision to concentrate all launches on to the STS, and actively to discourage the continuation of civilian ELV production in the USA. The vulnerability of this single-mindedness was exposed when the Shuttle was grounded after the Challenger accident, and the European Ariane launcher found itself virtually alone in the commercial ELV field.

The ESA Spacelab programme was similarly criticized in Europe as being a 1 billion US dollar gift to the US STS programme. It is true that all ESA received overtly in return was the free use of half of the payload bay on the first flight, but this was Europe's entrance ticket to manned space flight. Through the Spacelab programme ESA and large sections of European aerospace industry received their baptism in the, for them, quite new field of manned space flight.

But the breakneck pace of space development did not allow much time for quiet reflection on the technical success of Shuttle and Spacelab. Early in 1984, less than three years after the Shuttle's first flight, the US President announced that he had directed NASA to develop a permanently manned space station. Friends and allies were invited to participate in the development of this international space station. After another extended period of studies and negotiations (it took until 1989 to reach agreement), Europe, Japan, and Canada all undertook to participate in what is now called International Space Station Freedom. The basic Space Station design has undergone several significant changes over the past four or five years, and Congress has shown a yearly reluctance to vote as much funding as NASA has considered necessary. Nevertheless, 1990 seems to be the year in which the US programme can be said to have been given the green light, although for a rather smaller configuration than NASA had planned. Moreover, with the US system of annual budgetary votes, there is no possibility of guaranteeing a peaceful passage from now onwards. Minor changes will continue to be made, but the final configuration should look pretty much as in Fig. 3.4.

If all proceeds more or less as planned, Japan, Canada, and ESA will each have a module attached to the main space station. These modules will, to a certain extent have specialized facilities, and it is intended that there will be a good deal of exchange between the partners. Europe is not, however, satisfied with merely having a module attached to the US station, and ESA's long-term plan also calls for the almost simultaneous development of an autonomous module which would be visited from time to time by astronauts,

Fig. 3.4 Space station configuration

but be capable of automatic operation between visits—hence its name: the Columbus Free-Flying Laboratory (FFL).

On both sides of the Atlantic there has been considerable criticism over this significant expenditure on a space infrastructure for which the day-to-day use is by no means clear. The inevitable daily expense of operating and maintaining such a complex infrastructure makes it unlikely that it can ever become a commercially viable operation. A previous NASA Administrator, Dr James Fletcher, liked to say that the most important uses for the facilities of the Space Station are still unknown, and will only begin to emerge once the facility is available.

This is not place to attempt judgement: suffice it to say that the whole venture—both for the US and its partners—is to some extent an act of faith. That it will greatly stretch our knowledge and experience in many disciplines is beyond doubt. Whether it will produce the equivalent of a 21st century Gold Rush is far less certain. Middle of the road specialists are already beginning to say that it is the programme we cannot afford not to do. As the various elements start to be put into place by the Shuttle, the real value of this brave operation will gradually be revealed.

The excitement of the space stations should not, however, be allowed to obscure a more modest European programme which grew out of the Spacelab programme. It is called Eureca, short for European Recoverable Carrier, and will be designed to take a payload of about 1000 kg—unmanned, of course (Fig. 3.5). The plan is to launch it by the Shuttle, eject it, leave it in orbit for several months and then recover it with a later Shuttle launch. Eureca is designed to be re-used several times. Many analysts consider this programme as a more realistic way of proceeding for Europe, rather than attempt the expensive jump straight to manned space stations. Unfortunately, the poor Eureca does not have the glamour of a manned programme. Perhaps, however, its value will come to be recognized after the first flight which, subject to the Shuttle's launch schedule, should be in 1991/92.

Returning to the Soviet programme, Salyut was certainly not intended to be the Soviet Union's last word on space stations, and by 1986 they had replaced Salyut by the Mir space station. First heralded as a station to be permanently manned, Mir was in fact left vacant for a period in 1989, and many observers saw in this the evidence that attention was being paid to the Soviet critics of the expensive manned programme. Since that time, however, occupation of the station has been resumed, but insistent attempts to sell experimental space and even cosmonaut places indicate that the Soviet programme is under a pressure long known in Western space programmes: to produce some income.

Two more items are necessary to complete the panoply of elements for manned space flight: some form of motor to assist the astronaut to move about during EVA, and what we might call a space taxi, to take astronauts

Fig. 3.5 ESA's retrievable carrier—EURECA—during assembly

Fig. 3.6 NASA astronauts Gardner and Allen—equipped with back-pack Man Manipulator Units—bringing a stranded satellite back into the Shuttle for refurbishing on Earth and re-launch

and materials from one spacecraft to another, even when they are in significantly different orbits. The first of these exists already. The US Man Manoeuvring Unit (MMU) is a spectacular piece of equipment (Fig. 3.6) and has been used successfully on many occasions. It weighs about 148 kg and can be operated for up to about 6 hours at a stretch, but its range is limited to around 45 m from its spacecraft base.

In February 1990, a Soviet cosmonaut left the Mir space station for a test ride in space on what has been called both a 'space motorcycle' and a 'space armchair'*. In fact, it is not dissimilar to the US MMU, and demonstrates

* Officially known by the Soviets as SPK, an acronym from the Russian phrase 'means of forward movement for cosmonauts'.

the importance both space programmes attach to being able to have astronauts work in the area around space stations.

The second apparatus, the 'space taxi' (known as the Orbital Manoevring Vehicle) is already under development in the US, and one may suppose that the Soviet programme has something similar on the stocks.

As the space infrastructure becomes more elaborate, there is an increased need for more sophisticated handling equipment. The Canadians were quick to see an opportunity here for their industry, and their contribution to the STS programme was the development of a remote manipulating arm which is now attached to the cargo bay of the Shuttle (Fig. 3.7). Operated by astronauts within the Shuttle, Canadarm (as it is dubbed) has been successfully operated to lift satellites into the Shuttle's cargo bay. At the start of 1990, it was used to recover the Long Duration Exposure Facility (LDEF), a 10 tonne satellite, nine metres in length. LDEF, which carried 57 scientific experiments to test the effect on materials of long exposure to space, was launched in 1984 and was intended to be recovered a year later. The Challenger Shuttle accident, of course, upset all NASA's plans for using the Shuttle, and LDEF had to wait. After more than five years in space, the spacecraft's orbit had gradually degraded, and it would have fallen back to Earth only weeks after the date of its spectacular recovery in January 1990. It will take some time to analyse completely the effect of this unintended long stay in space, but a first examination showed that although the outer skins had been dented by more than 10 000 collisions with micro-meteorites, the aluminium struts forming the main body were still intact.

The Canadian element in Space Station Freedom will be equipped with an even more sophisticated version of the manipulating arm, and this is an indication of the shape of things to come. Not only will many of the station's structural elements need to be manhandled out of the Shuttle and into position, many specialists believe that a sort of repair garage is needed where spacecraft can be repaired, have their fuel topped-up, films and tapes exchanged for new ones, and even whole subsystems up-dated with models using the latest technology.

For completeness, mention must be made of the astronauts' suits. These are not so simple as they may seem, for they have to allow the astronaut to live, breathe, and work in the maximum degree of comfort and safety—including some protection against the inevitable space bombardment of small particles. Until recently the manufacture of these suits (which are not, of course, worn in the pressurized laboratories and crew quarters on the Shuttle and the space stations) was a monopoly of one specialist manufacturer in the US and, one must assume, another in the USSR. It is a sign of the growing number of manned programmes that Europe too plans to develop the art of producing these suits. A European industrial consortium has been formed to design two new spacesuits: one for use inside space stations (IVA—

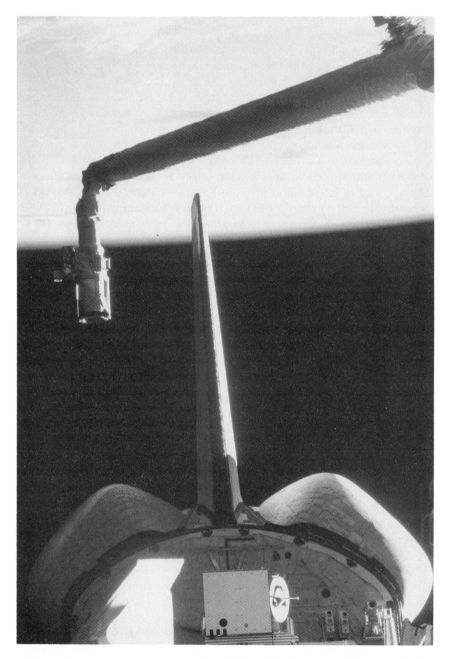

Fig. 3.7 A view of the open cargo bay of the Shuttle, with the Canadarm being deployed to lift out the waiting satellite, and leave it to make its own way to final orbit

intra-vehicular activities) and the other for EVA (extra-vehicular activities). This is no simple task: a sum of over 10 million US dollars is to be spent on the design phase alone. Technological independence does not come cheap.

And so, less than 35 years after the launch of the first, football-sized satellite, manned space flight has become almost routine. Instead of cautious, single Earth orbit ventures of instantly world famous astronauts and cosmonauts, we have already graduated to more or less permanent structures in space, where scores, or even hundreds of astronauts will live and work. It is only natural that such progress stimulates fertile imaginations (of respected scientists and engineers, as well as of talented dreamers) to predict holiday space tours and even permanent and more or less self-supporting colonies in space. Tickets are said to have already been sold for space vacations and, more gruesomely, for the privilege of having one's ashes encapsulated and placed into an eternal orbit. It is tempting to dismiss all this as coming from the lunatic fringe, but the reputations of some of the proponents—not perhaps those sponsoring space burial—are such that one is bound to take note of the marked trend towards a rapidly increasing human population in space, and an equally rapid increase in the size and complexity of the structures there. Who would dare to prophesy what the scene will be in another 35 years?

USES OF SPACECRAFT

After reading about the furious activity in the development and production of launch vehicles, the reader is entitled to ask what sort of things are, in fact, waiting in such numbers to be put into space. There are, crudely put, four major reasons for wanting to launch a spacecraft:*

(1) to look outwards, without the handicap of the Earth's atmosphere, and so to be better able to observe planets, stars and comets and other galaxies;

(2) to look backwards towards the Earth in order to observe either the Earth's atmosphere, or the surface of the Earth itself;

(3) to use the spacecraft as a relay station off which signals can be bounced back to one or more places on Earth; and

(4) to carry out experiments—and perhaps even production—without the inconvenience of the Earth's gravitational force.

All the present uses of spacecraft are exploiting one or more of these four

---

* 'Satellite' is reserved for a body put into orbit around the Earth. 'Spacecraft' includes all satellites, but also all other bodies which are sent outside the Earth's gravitational pull.

advantages. It is important, too, to add that the use of a spacecraft must either enable something to be done which has previously been impossible (one can cite the Hubble Space Telescope as an example), or allow it to be done cheaper, easier, or better than has otherwise been possible.

In examining the uses of space, it is proposed to follow approximately the order in which the uses have been developed, starting with space science including microgravity, and passing through telecommunications to reach meteorology and Earth observation.

4. Space science

At the risk of offending purists and specialists alike, space science in this section will be restricted only to the science of space, that is to say of the Earth, the planets, and what lies around, between, and beyond them. It will not include the science which can be conducted in space either by automatic or manned spacecraft, such as investigations using the microgravity environment in a spacecraft.

It cannot be too often repeated that the advances achieved in space techniques owe much to the vision and demands of space scientists. The earliest sounding rockets and satellites contained only scientific experiments. The 1960s and 1970s were thus a period of tremendous space science activity, with fruitful cooperation between large numbers of scientific groups in most parts of the world. Of course, there was also intense rivalry, and the familiar striving to be first to publish new and exciting results, but the irritations caused by such human foibles were massively out-weighed by the genuine international cooperation.

The space scientists often pitched their requirements too high for the engineers to be able to meet their specifications fully, but this proved a constant incentive to spacecraft and instrument designers, without which the technological progress would not have been so rapid. The scientists' ever-increasing need for a means of transmitting data to the Earth opened the way to the development of commercial telecommunication satellites. The requirement for pointing accuracy in space-borne optical instruments similarly advanced the state of the art, and virtually every type of scientific sensor or counter has proved useful in what we now call 'application' satellites, i.e. those which have a purpose other than scientific research. Even today, aerospace firms are keen to win contracts for scientific satellites, because they know that they will be required to work on the frontier of the technologically possible—and even beyond it, if the space scientists had their choice.

The very close working relations between the engineers and scientists in the space programmes and the optical astronomers, and the mutual respect which was built up over the years, have considerably enriched both communities.

Many graphic illustrations have been coined to show the value of space science experiments over the past thirty years, but quantification is of doubtful value and indeed only incites ground-based scientists to speculate on the discoveries they might have produced, if they had had access to the same amount of funding. This is a long-standing quarrel in which more-or-less innocent bystanders can easily be trampled underfoot. The safe line to adopt is to see space science as a valuable contribution in many scientific disciplines. This much is incontrovertible, and it has the merit of veering very much on the side of modesty.

The ESA Science Directorate has used the following classification in developing its long-term scientific programmes:

- astronomy;
- solar and heliospheric* physics;
- space plasma physics; and
- planetary research.

Let us borrow this classification for the purpose of seeing the relevance of the progress already made, and where the next big thrusts are likely to be.

Astronomy

Most information about the universe is obtained from interpreting signals in some part of the electromagnetic spectrum. Each part of the spectrum is characterized by a different frequency and wavelength of the electromagnetic wave or photon, all of which travel at the speed of light†. For many centuries the only wavelength region accessible for astronomical observations from the ground was the narrow part visible to the human eye. This ranges from 400 to 800 nanometres‡, a dynamic range of energy of only a factor of 2. Added to this, the visibility even from the most favourably situated observatories is limited by the Earth's atmosphere which obscures, bends, or defocuses the incoming rays of light.

Some 40 years ago, it was established that observations in the radio part of the spectrum (wavelengths of roughly from a centimetre to a metre) gave promising results. This led to the establishment of radio telescopes in several parts of the world. With the advent of spacecraft, operating above the Earth's atmosphere became possible for astronomical observations from long wavelength photons to short wavelength gamma rays. In this viewers were able

---

* *Heliosphere*: the region around the Sun where the influence of the solar wind is noticeable.
† 3×10^8 m/s, i.e. $3 \times 100\,000\,000$ m/s.
‡ 1 nanometre $= 10^{-9}$ m.

to obtain signals from a range around 75 times greater than that accessible in the visible region.

Discoveries made during the relatively short space era have drastically altered our concept of the universe and have led to the discovery of many entirely new cosmic phenomena and objects. Space observations are particularly valuable, because progress in astronomy and astrophysics (the branch of astronomy dealing with the chemical and physical constitution of the celestial bodies) often requires correlations between the observed characteristics in widely different wavelengths, as a means of interpreting the underlying physical properties.

A recent list of outstanding needs in modern astronomy and astrophysics included information related to:

- the formation and evolution of stars and planets;
- the structure and dynamics of the interstellar medium, i.e. that which exists between the star and the planets;
- the origin of cosmic rays;
- the dynamic and chemical evolution of stellar populations; and
- the large scale structure and evolution of the Universe.

From this it can be seen that, great though successes have been, much still remains to be done.

The International Ultraviolet Explorer (IUE) satellite programme is an excellent example of international ultraviolet astronomy cooperation, and a few details of the satellite itself and the way it has been operated will help the reader to understand how scientists put satellites to the service of their scientific disciplines.

IUE is a joint project between ESA, NASA, and the UK Science and Engineering Research Council (SERC). Launched into geosynchronous orbit in January 1978, the satellite has had over eleven years of useful life and, at the time of writing, was still operational (Fig. 4.1). It is equipped with a 45 cm telescope feeding into spectrographs sensitive to radiation in the ultraviolet region (in fact, in two ranges between 0.115 μm and 0.32 μm). The project was designed to provide a facility for ultraviolet spectrophotometry of sources of astrophysical interest, and two ground observatories were established to control the satellite. One, operated by NASA from the Goddard Space Flight Centre (named after the same Goddard who started his one man rocket crusade back in the 1920s) in Maryland, USA, had control of the satellite for 16 hours each day, and the other near Madrid, Spain, operated by ESA, took over control of the satellite for the remaining 8 hours.

Each space agency allocated observing time on the satellite to observers— just as is done with ground-based telescopes—after consulting their scientific advisory committees. Demand exceeded availability by a factor of more than

Fig. 4.1 Modern astronomers examining the images sent back to Earth by the successful IUE satellite

two, and IUE became the most productive of telescopes. Over 1600 scientific papers based on IUE data have been published so far in reputable astrophysical journals.

A special feature of the IUE programme is the rapidity with which processed data can be made available to the experimenter—normally within 24 hours. To encourage the experimenter not to sit on the data for too long, all IUE data become available to the public six months after the observation. In addition, 'national hosts' have been established in nine or more countries, from which experimenters can obtain access to archived IUE data. It was realized at an early stage that the way to get value for money in such a programme is not to invest everything in the satellite itself, but to be sure to be able to manipulate, archive and distribute the data using the most modern facilities of information technology.

Understandably, IUE is very popular with observers. Sitting in front of computer consoles in an air-conditioned room is perhaps not so romantic as the traditional picture of the astronomer huddled in blankets next to his telescope and swigging hot cocoa, but the results are spectacular. And, of course, IUE is just one of many such spacecraft which have been launched and successfully operated. The astronomical community had been waiting for several years for the Hubble* Space Telescope (HST) (Fig. 4.2), a joint NASA/ESA programme with its 2-metre-class astronomical telescope and associated instrumentation. Delayed partly because of the Challenger Shuttle accident and perhaps also because of its own complexity, it was successfully launched in 1990.

Like IUE, it is designed to be used as an observatory with the main science institute in Baltimore, USA. The call for experiment proposals was over-subscribed six times.

Journalists have tried to explain the importance of the HST by claiming that it would enable us to see back to the origin of universe. This is, alas, somewhat exaggerated, but it could certainly lift a number of veils, and its highly sensitive instruments, such as the Faint Object Camera, will allow the capturing of images which was never possible before. An important feature of the HST is its modular design, which simply means that one or other of its constituent parts can be removed or replaced without the whole spacecraft having to be taken apart. The Space Shuttle is intended to visit the HST approximately every three years, and astronauts will be able to repair, replace or even update not only the scientific instruments but also many of the spacecraft's critical sub-systems. In this way a total operating lifetime of 15 years is planned.

Unfortunately, however, the Hubble telescope, once in orbit, showed an

---

* Edwin P. Hubble (1889–1953), the US astronomer famous for his research using the Mount Palomar telescope.

Fig. 4.2   The mammoth HUBBLE space telescope in final testing

important flaw which significantly reduces the clarity of the images. A panel
investigating the cause of the fault has tentatively identified an error in one
of the devices used on the ground to test the mirror before flight. Efforts are,
of course, being made to devise procedures which will at least reduce the
effects of the flaw, but it is virtually certain that nothing near to the expected
results will be achieved until physical repairs have been effected on the

spacecraft. Such an operation is expensive and needs lengthy preparation, but there are hopes that it can be undertaken in 1993. Meanwhile, in spite of its handicap, Hubble is producing some exciting results.

ESA's Hipparcos satellite is another example of salvaging important scientific results from a project marred by hardware failure. Successfully launched by Ariane in August 1989, the apogee boost-motor attached to the satellite failed to ignite, leaving the satellite in a highly elliptical orbit instead of the planned geostationary orbit. Nevertheless, with a lot of behind-the-scenes ingenuity, it is proving possible for Hipparcos to achieve its mission: to gather information on the position, motion, and distance from the Earth of 120 000 stars, and all this with a greater precision than ever before. The accuracy of Hipparcos is the same as that needed to focus from Earth on a man standing on the Moon. The operation will take longer than planned because of the faulty orbit, but the satellite appears in sufficiently good health to be able to complete this new celestial atlas.

Solar and heliospheric physics

The Sun forms the centre of the solar system, and is its principal source of energy. Solar and heliospheric physics is the study of the Sun and the heliosphere, and their interplay with the planets. The Sun is the only star whose structural features can be directly detected, and it thus provides a valuable source of ideas about the composition and behaviour of other stars in the Universe.

The Sun and the solar system are in fact now used as a gigantic laboratory. Of particular importance is the study of the magnetic fields and their interaction with plasmas; it is now understood, for example, that the origin of the Sun's coronal* heating is to be found in the magnetic field. Another fundamental area of research, which is considerably aided by spacecraft data, is the effect of the solar transients: shocks propagating through the solar winds.

ESA's Ulysses spacecraft, launched from the US Shuttle in the autumn of 1990, is an especially interesting cooperative programme in this field. It will make a five year journey to fly over the hitherto uncharted poles of the Sun. To achieve this, the spacecraft needs more thrust than can be presently provided by launch vehicles, and so it is aimed to pass near the planet Jupiter and to be slung into the desired orbit by the planet's gravitational force (Fig. 4.3).

Previous satellites have all been in orbits in the plane in which the planets

---

* The corona is the rare gas halo surrounding the Sun.

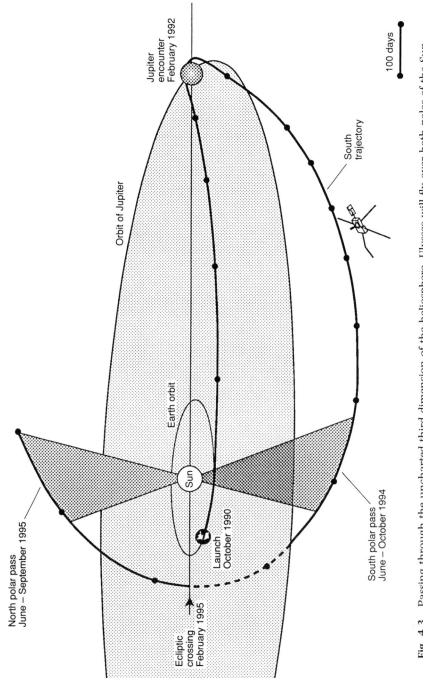

Jupiter
encounter
February 1992

Orbit of Jupiter

South
trajectory

100 days

Earth orbit

Sun

North polar pass
June – September 1995

Launch
October 1990

South polar pass
June – October 1994

Ecliptic
crossing
February 1995

Fig. 4.3 Passing through the uncharted third dimension of the heliosphere, Ulysses will fly over both poles of the Sun

orbit the Sun, the so-called 'ecliptic' plane. Ulysses will be the first spacecraft to venture into the 'out of ecliptic' plane. Its nine scientific instrument packages will yield valuable data on solar wind, solar flare X-rays, magnetic fields, and a host of related areas of prime scientific interest.

It is to some extent misleading to be dealing with the various categories of space science as though they were watertight areas of research; there is, of course, much interaction between them. The study of solar and heliospheric physics is, for example, not only germane to basic astrophysics, but also to such areas as plasma physics, atomic and particle physics, and magneto-hydrodynamics.

A good example of this interdependence is the Solar–Terrestrial Programme (STSP) which was approved in 1986 as a cooperative venture between ESA and NASA. This programme consists of two separate projects, Soho and Cluster—each has its own scientific objectives, but the total value is greatly enhanced by the possibility of comparing the data simultaneously acquired by the five spacecraft: Soho and the five smaller spacecraft comprising the Cluster project. These latter are described below under the heading 'Space plasma physics'.

ESA's Soho—the projected Solar and Heliospheric Observatory—is due for launch in 1995 and is designed to cater for the wishes and aspirations of experimenters in this important area of space science. It is intended to investigate:

(1) the physical processes which form and heat the solar corona and produce the expanding solar wind; and

(2) the solar interior structure, using the methods of helioseismology and by observance of solar irradiance variations.

The payload of the Soho spacecraft will weigh around 650 kg and will consist of twelve experiments, nine designed by European and three by US scientific groups. The whole payload will be an integrated package requiring coordinated operation and data analysis between the scientific groups, and will have a design lifetime of two years. An Experimental Operations Faculty, again to be located at NASA's Goddard Space Flight Centre, will be the focal point for these coordination activities. Its main task will be to organize the operation of the payload and to control the solar remote sensing imaging and spectrometric instruments during the daily contacts between the spacecraft and the ground. Control is provided by ground stations belonging to NASA's Deep Space Network (DSN) because the so-called 'halo' orbit is too far from the Earth to be reached by the ESA ground stations. (The spacecraft will in fact be in orbit around a point one and half a million kilometres from the Earth* which has been specially chosen to suit the kind of observations to

---

* The L1 Sun–Earth Lagrangian point: the gravitational saddle point between Sun and Earth. One of several such astronomical milestones identified by the French mathematician Lagrange.

be made.) There will be daily contacts of three short (1.3 hours) and one long (8 hours) periods; scientific data acquired at other times will be stored on magnetic tapes on-board the spacecraft and transmitted to the Earth during the contact periods.

These next years will be busy ones for all involved in the Soho project. A Science Working Team, formed of the Principal Investigators for each experiment, set the scientific requirements for the satellite's development and created a number of working groups to deal with specific topics, such as the Solar Corona and Particles Working Group, which is already starting to define how Soho will be operated and what ground facilities will be needed to achieve the project's objectives.

Space plasma physics

The spacecraft which have probed into space over the past 30 years have shown that it is filled with dynamic and energetic plasma: mixtures of charged particles of various densities. Plasma in the solar system has a cellular structure and thin boundary layers separate plasma regimes which have markedly different characteristics. For example, the magnetopause is the boundary which separates the magnetosphere—the magnetic field systems surrounding the Earth—from the plasma and magnetic field of the solar wind—the extension of the Sun's atmosphere. Instruments aboard spacecraft have provided proof of a previously unimagined complexity in the interaction between electromagnetic fields and charged particles. European scientists recently reviewing the progress that has been made*, pointed to two particularly surprising observations. Firstly, although the ideas of pioneers in this field successfully described the basic features, *in situ* measurements in each newly explored region have each time presented major surprises. The second observation is that in spite of the diversity of plasma environments in the solar system, the same plasma processes are often found to play a dominant role. Although there can never be enough experimental data to satisfy the scientists, specialists in this field believe that the research is sufficiently mature to allow a quantitative study of plasma processes, rather than to remain at the stage of collecting further samples of interesting phenomena.

* Report of ESA's Topical Team on Space Plasma Physics ESA SP-1070 (1984).

Table 4.1 Data on the nine planets of the solar system

	Diameter (km)	Average distance from Sun (million km)	Moons
Mercury	4850.00	57.6	None
Venus	12032.00	107.5	None
Earth	12739.00	148.0	1
Mars	6755.00	225.0	2
Jupiter	141968.00	773.0	16
Saturn	119296.00	1413.0	At least 20
Uranus	52096.00	2583.0	15
Neptune	49400.00	4500.0	8
Pluto	3040.00	5865.0*	1

* Pluto's orbit around the Sun is highly elliptical; its closest point to the Sun is 4384 million km. Neptune is able to claim to be the farthest planet from the Sun when Pluto's orbit cuts inside Neptune's orbit. This is indeed the situation between 1979 and 1999. Thereafter, Pluto will reign for the next 228 years.

Planetary research

Continuation of the age-long search for information about the planets perhaps needs no explanation or justification. Almost all we know about the planets has been gained from space probes over the past 25–30 years. By 1990 unmanned spacecraft had either skimmed, flown by, orbited, or landed on all but Pluto, and the vast amount of data and images have confirmed some theoretical predictions or upset or modified others, and posed a whole new set of problems for our scientists.

Obviously, a great deal of research is motivated by a desire to know in more detail how the solar system developed, and to gain a better understanding of the processes involved. Given our increased chances of making spot surveys, there is also an interest in looking for accessible sources of new materials to supplement the Earth's dwindling resources. Also present—and always of interest to the media—is the possibility of discovering other forms of life. All this adds up to a rather heady mixture, and planetary programmes are generally real 'space spectaculars'.

In the USA, NASA started its long series of planetary missions with Mariner 2, which flew close to Venus in 1962. In the remainder of the decade, five more Mariner missions re-visited Venus and made flights close to Mars. The

USSR had a similar programme and from 1965 the Venera series of spacecraft (there were 16 of them over the next 18 years) provided fly-bys of Venus, landings, and atmosphere entry probes. A parallel series—Mars 2–9—performed the same amazing feats on the planet Mars.

The last of the Mariner series, No. 10, in 1974 achieved not only another successful fly-by of Venus, but the only fly-by there has so far been of Mercury.

1976 brought a new sensation: Viking 1 and 2 made soft landings on Mars, and the 'rover' sent back dramatic data (Fig. 4.4). The Soviets in this period gave Mars a rest, and until the mid-1980s, appeared to be concentrating on Venus, including a new series of spacecraft, Vega 1 and 2.

NASA too, devoted a new series to Venus in 1978 with Pioneer 1 and 2, but the prize for the most complicated planetary mission so far must go to the US Voyager 2. Launched in August 1977, it was intended to take advantage of an alignment of the planets which takes place only once in 176 years. Its fascinating itinerary reads as follows: Jupiter—8 July 1979; Saturn—26 August 1981; Uranus—27 January 1986; and Neptune—24 August 1986, after which the spacecraft headed into eternity. Figure 4.5 is a typical product of this mission.

When assessing the extent of this success, it is good to remember that Voyager's transmitters have about the same power as a very weak electric light bulb—22 watts. The signal from Neptune takes over four hours to arrive on Earth, and by this time it is 200 million billion times weaker. Over 30 powerful receivers on Earth—three with a diameter of about 70 m and new, sophisticated listening equipment—were used to listen for Voyager's signals.

In recent years attention has swung back to the planet Mars. The Soviets, with their not altogether successful Phobos, rekindled interest in 1988–89, and have since that time been insistently urging the international space community to band together for a Mars mission. The campaign gained respectability in the USA when President Bush included reference to a mission to Mars 'and beyond' in a speed in mid-1989 on the future of the US civilian space programme.

There is enormous interest from space scientific groups all over the world, and there is no shortage of offers to develop new and exciting instruments, but without the major participation of one, or preferably both, of the space giants, there will not be enough funding for such ambitious programmes as are being touted by the enthusiasts. Whether they will be able to afford this as well as the already committed space station programmes seems doubtful, and this, of course, explains the hostility of parts of the space community to this enormous expenditure on space infrastructure.

Meanwhile, ESA has decided to cooperate with NASA in the Cassini mission, which will put a spacecraft in orbit around Saturn for a four-year

Fig. 4.4 The surface of Mars as captured by the Mars lander during NASA's VIKING project

Fig. 4.5 Jupiter and its tiny satellites Io, Europa and Callisto as seen by NASA's Voyager in 1979

observation period. ESA will provide a sound (known as Huygens) which will be dropped into the atmosphere of Titan, one of Saturn's satellites (Fig. 4.6). The data gathered by Huygens will be relayed to the mother spacecraft and re-transmitted to Earth.

Another active branch of this part of space science concerns comets. Scientists have long been interested in comets because they are thought to have been formed at the time of the birth of the solar system, some four-and-a-half billion years ago. Space programmes have produced a number of interesting cometary observations, but the most important was

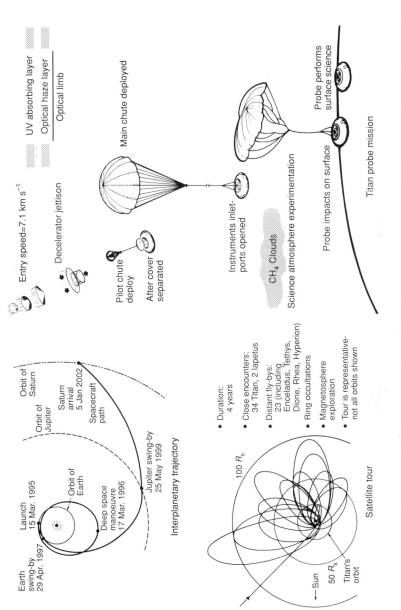

Fig. 4.6 The Cassini spacecraft must undertake a series of complicated manoeuvres over a 6–7 year period in order to perform the various parts of its mission: an asteroid fly-by, Jupiter fly-by and a tour of the Saturn system

Space science

the rendezvous with Halley's comet in 1986. The comet comes by only every 76 years and so this can accurately be called a chance in a lifetime, and it was seized by an international group representing ESA, NASA, the Soviet Union, and Japan.

Each agreed to contribute observations from spacecraft to be launched at times, and in orbits, which would provide a succession of reports to enable the final rendezvous to be made with precision. All the spacecraft provided valuable data on the comet, but the climax was provided by ESA's Giotto spacecraft, which gave images taken as close as 600 km (Fig. 4.7).

Altogether, 50 scientific experiments were directed at the comet, and more than 500 scientists from 100 institutes all over the world took part in the

Fig. 4.7   The image of the Halley Comet transmitted by ESA's GIOTTO satellite

programme. Beyond the impressive 'pictures' of Halley's comet, the wealth of data obtained will require years of patient analysis.

Encouraged by this success, space scientists are examining a project to meet another passing comet, and to obtain samples of its nucleus. Whether this project will survive the ordeal of examination by the scientific—and financial—committees, remains to be seen.

The most expensive of the space sciences, planetary research has bounced back into favour and funding. At the end of 1989 the NASA spacecraft, Galileo, was launched towards Jupiter, where, in December 1995, a probe will be released into the Jovian atmosphere and the spacecraft will orbit the planet for 22 months. If all goes well, this will provide another valuable piece in the complicated puzzle of the solar system.

A remarkable feature of Galileo is the use of three planet passes to acquire the additional momentum needed for the voyage to Jupiter. A few months after launch, the spacecraft rounded Venus and thereby increased its speed by 5000 mph (8000 km/h). The spacecraft changed direction and headed back to Earth for its second and third flybys. Only then will it be ready for the trip to Jupiter.

The Magellan spacecraft is another example of recent planetary research. Designed to spend eight months in orbit around Venus, the spacecraft, which arrived in orbit in August 1990, is expected to map about 90 per cent of the planet's surface. Scientists are hopeful that thereafter it will be possible to manoeuvre the spacecraft into a position where it will be able to continue doing useful work.

It is now time to move on to the Cluster project, referred to above as part of the STSP. It provides a good example of space science projects and, apart from the ESA/NASA cooperation, it gives another opportunity to work with Soviet space scientists. The project consists of four spacecraft built to extremely stringent requirements of electromagnetic cleanliness, so that their instruments will be able to make accurate electric field and plasma cold measurements. This mission has been designed primarily to study the three-dimensional formation and behaviour of small-scale structures in the Earth's plasma environment—'small', that is, on an astronomical scale: in fact, a few thousand kilometres in length! The aim is to study the transfer of mass, momentum, and energy across boundaries of different plasma regions, and to obtain a much more precise understanding of the processes involved.

The process of having one's scientific payload flown in a spacecraft is long and complicated. The costs of satellite, experiment, launcher, and the subject acquisition and treatment of data put the enterprise beyond the reach of most individual laboratories. The prime movers are the space agencies, such as ESA and NASA, which have long-range space science programmes, well known to the space science community. Nowadays the agencies' programmes are not only designed nationally or regionally to give some satisfaction to each of the different classes of space science, they are also coordinated between the agencies with the aim of reducing overlap. A wider coordination is often achieved through such international organizations as COSPAR, the Committee on Space Research, which has been active since 1958. COSPAR is a committee with a charter under the International Council of Scientific Unions (ICSU), and it is proof of the multi-disciplinary nature of space research that 13 of the present 18 international unions within ICSU actively participate in the work of COSPAR. Such international committees and unions are tremendously influential in determining which major space research missions are flown, for their senior members are consulted by the space agencies, and at home they invariably sit on the national committee which decides on the allocation of funds to space science groups. Efforts are sometimes concentrated internationally by such devices as the International Geophysical Year (1957–1958), the Solar Maximum Year (August 1979–February 1981), and the International Space Year in 1992. Small scientific groups need to be very mindful of these events when preparing their proposals for experiments. It is in no way contrary to scientific integrity for a head of laboratory to assess the chances of obtaining national or international funding before setting his laboratory on the long haul. And long it is; many scientists will have been associated, for example, with the IUE (International Ultraviolet Explorer) for well over twenty years. It is not unusual for 10 or 12 years to pass between the first discussion of a new scientific experiment and the date of a successful launch. In fact, from the first studies of the Ulysses programme to the passage of the poles of the Sun, 25 years will have elapsed.

The relatively few launch opportunities and the large number of experimental groups throughout the world account for the feverish activity whenever a space agency invites applications to take part in a new satellite project. In practice, by the time the official announcement is made, most of the premier groups will already have formed their alliances and started to prepare for the wearying process of evaluation and selection—or refusal.

When the scientific mission has been declared feasible, it is usual—at least in the ESA—to award what is known as a Phase B industrial study (and sometimes two studies in parallel as a means of producing the best possible technical solutions to the problems posed by the scientists' requirements). If this study confirms that the satellite can indeed be built and operated more or less as planned (and during this phase the scientists often have to reduce their requirements in order to keep the cost and schedule within reasonable bounds), the final specifications are written and industry is invited to make costed proposals for the Phase C–D contract—the development and production of the satellite.

In some programmes the scientific experiment is also built by industry as part of the overall satellite contract. This is usually the case with satellites which have a single large instrument, such as a telescope, as payload. Where the satellite carries a collection of smaller experiments from different laboratories, however, it is the practice in the ESA to oblige these laboratories to seek their own, national, source of funding. In spite of all these hurdles, there is no diminution in enthusiasm, and the drive and initiative of our space scientists have produced benefits way outside what many still consider to be the esoteric realms of space science.

The search for extraterrestrial intelligence (SETI)

It is temptingly easy to mock those who seek intelligent life elsewhere than on Earth, and taking them seriously has not been helped by the spate of films on the subject. Nevertheless, it should be remembered that there is a significant group of distinguished scientists—from most parts of the world—who dedicate their professional life to the study of the subject.

NASA spent a lot of time and effort on SETI around 1977 and came to the conclusion that: 'it is both timely and feasible to begin a serious search for extraterrestrial intelligence'. The NASA report recommended that 'modest resources' be made available for a SETI programme.

The activity is, by its very nature, international in character and for that reason, the SETI activities of the International Academy of Astronautics (IAA) is particularly important. Two sessions are regularly devoted to SETI at the annual congress organized by the International Astronautical Federation (IAF) to review SETI projects, as well as to discuss non-technological aspects including even the legal, social and political aspects of SETI.

The discovery of other intelligent species by detecting their radio transmissions is the basis for most serious SETI projects. This was first advocated by Cocconi and Morrison in 1959, when they proposed that a search be made in the relatively quiet frequencies near the wavelength of 21 cm emitted by hydrogen. Since that time many dozens of searches have been carried out, and there is no shortage of proposals for new areas of search. No sign of extraterrestrial life has yet been detected, but those dedicated to this rather specialized area of space science point out that SETI is still in its infancy. They also continue to plead for more coordinated international projects.

It must be stressed that the work has generally been characterized by the same scientific rigour one has come to expect of space science programmes, and the participants are well aware of their vulnerability to false alarms and even hoaxes. Validation procedures must play an important part in any future SETI projects.

Although in these days of telecommunications satellites, earth observation satellites, and other practical space applications, the SETI scientists cannot claim to be in the mainstream of space development, the search for other life must certainly be counted as a legitimate and worthwhile activity. The question as to whether other intelligent life exists has concerned philosophers for centuries, and it is understandable that one section of the space community should feel called to continue the search with the newly available facilities. The consequence of a successful identification of extraterrestrial life would certainly provide a massive justification for the effort.

5. Microgravity

Microgravity is the state of near-weightlessness that exists within a spacecraft which is in orbit around the Earth or has achieved escape velocity and is on its way towards other planets. The gravity on Earth is used as the standard measurement—1 g—and the gravity experienced in a spacecraft is expressed as a fraction of that, for example, 10^{-6} g, i.e. one millionth of the gravity on Earth. Inversely, the pressure to which astronauts are subjected during lift-off (hypergravity) is expressed in multiples of g: 3, 4, 5 g, etc.

In the microgravity environment gravity-induced convection and the sedimentation processes are suppressed (the extent depends on the degree of microgravity achieved), and hydrostatic pressure disappears. In such an environment certain physical phenomena familiar on Earth are radically altered. For example, matter tends to levitate and the behaviour of fluids becomes quite different. Realization of this fundamental change had led to the development of what is loosely termed 'microgravity research'. In fact, it means experiments conducted in a microgravity environment to gain a better understanding of the physical processes through an examination free of the disturbing force of gravity (Fig. 5.1). Microgravity research also includes, though at a very early stage, manufacturing materials of a composition which would be difficult, expensive or even impossible on Earth because of the presence of gravitational force.

As we shall see when we come to space commercialization in Chapter 11, a microgravity activity is sometimes started with the intention of manufacturing large quantities of a product, but experimentation in space quickly gives a deeper understanding of the underlying physical processes which enables the product to be manufactured on Earth.

Many different areas of microgravity research exist, and the aims and procedures vary sufficiently to justify a brief explanation of the main ones.

Materials science

The most striking work is in the production of single crystals of a size and purity unobtainable on Earth. This development is of considerable interest

Fig. 5.1 Inside SPACE-LAB—ESA astronaut Wubbo Ockels and NASA astronaut are themselves wired-up for experiments. While Wubbo concentrates on his 'weight-less' egg, an empty sock seems to be menacing his col-league

to the electronics industry. It has also been possible to grow large protein crystals, large enough to allow their structure to be determined through X-ray analysis. The higher growth rate and the final size of space-grown crystals could provide a new opening in biochemistry.

Fluid sciences

As with solids, the experiments can be made without the liquid being enclosed in a container, and the absence of important disturbing perturbations simplifies both the conduct of the experiment and the subsequent analysis of results.

Space biology and space medicine

Certain biological processes in animals and plants are sensitive to gravity, and the presence of gravity on Earth has made it difficult to study these sensitivities. How gravity is sensed, what the limits of sensitivity are, and the mechanisms of response developed in living systems are all problems which are not yet fully understood. The interesting results which have been so far achieved are not all explainable in terms of current biological knowledge. Our understanding of human physiology can also benefit from microgravity research. Many of the experiments have been oriented to ensuring the health of astronauts: understanding how the cardiovascular system adapts to weightlessness, and how bone and muscle develop with time are just two of a host of important problems. Perhaps of even greater practical value, however, is the realization that much of this space-borne research can be of direct relevance to terrestrial medicine, and not simply for the benefit of astronauts.

Because of the high cost of flying experiments in space, and also to get a reasonably cheap preview of the functioning of an experiment designed for operation in a space station, scientists are anxious to find cheaper and simpler ways of achieving near weightlessness. There are currently at least three methods in use:

1. Sounding rockets (see Chapter 1) are used to produce microgravity conditions for up to about six minutes, during which time valuable observations can be made.

2. Drop towers or converted mineshafts provide a cost effective way of achieving weightlessness for just a few seconds, and this is sometimes

adequate for the scientist's purposes, particularly in the early stages of research* and

3. Parabolic flights in specially-equipped aircraft can produce weightlessness for around 25 seconds. Provided that the experimenter has a strong constitution, a three-hour flight can easily produce 25 parabolic opportunities for experimentation.

Experiments in all the various disciplines which are beginning to show an interest in microgravity research need more than simply the microgravity environment; they need facilities and equipment. Skylab, Spacelab, and the US and Soviet space stations all require more or less complicated facilities, varying from furnaces to devices dedicated to the growth of protein crystals, or elaborate sledges which catapult the astronaut the length of the laboratory. Many of these facilities are extremely costly—in the millions or even tens of millions of pounds—but they are, of course, intended to be used by many experimenters.

In most countries it is accepted that money should be spent on microgravity research, and there are flourishing programmes in many, so far as the West is concerned, in preparation for the availability of Space Station Freedom. Indeed, some of the justification for developing the space station came from the claims that were made for microgravity research. It is evident that some, at least, of these claims were overly optimistic. Indeed, in the UK, the official view is that microgravity research should have a very low priority compared to other fields of scientific research which are already short of funds. The vigorous German programme seems also likely to be somewhat reduced in the 1990s. There is, however, a risk that this view is a rationalization born of the general shortage of research money, and those working in the microgravity field in the UK are in no doubt that further and substantial benefits will accrue. It is fair to say, however, that the way to successful microgravity research passes through solid, ground-based scientific research. Space-borne microgravity activities, with the exception of eventual manufacturing projects, are an extension of ground-based research, rather than a substitute.

As in so much of space work, the whole picture could suddenly be changed by the arrival of brilliant new ideas, and this is why it is so necessary to encourage some of our brightest young scientists to choose this path, in spite of the deprecating attitude of some of their peers.

It is, however, only fair to report that for some time a number of leading microgravity scientists have been turning at least part of their attention to the possibility of using hypergravity, i.e. under a gravitational force greater

---

* In 1785, William Watts of Bristol built a hollow tower in which to drop molten lead which solidified on the way down and produced higher quality lead shot than otherwise possible.

rather than lesser than normally prevails on Earth. There is some preliminary evidence that experiments placed in the same giant centrifuge which figures in astronaut training are giving interesting results. Even so, the better understanding of the physical processes involved undoubtedly stems from experiments in microgravity conditions. In other words, the jury is still out and the final verdict cannot be expected for many a long year.

6. Telecommunications

Nowadays people often do not know whether their telephone call or television picture owes anything to satellite technology or not. Provided the service is satisfactory in quality and price, the user is entitled to be indifferent about the means. The situation is, however, by no means static. The possibilities for using communications satellites, 'satcoms' as they are called, are increasing with development of the technologies, but also the telecommunication, information, and broadcasting communities have become more aware of the extent to which satellites can help them to provide new or better and cheaper services. At the same time, conventional telecommunications technologies have made major advances—in the development of fibre optic cables, for example—and service providers are obliged to look very carefully into the economics of new systems.

The specific idea of using a satellite to relay radio signals around the Earth dates back to 1945, when Arthur C. Clarke described it in an article published in the British journal *Wireless World*. The article accurately described the energy needed to put a satellite into an orbit approximately 36 000 km above the equator, at which distance it would move round at the same angular rate as the Earth and so appear to an observer to be in a fixed position. This has come to be known as the geostationary orbit (see Chapter 1).

Although the specific idea of using a satellite in GEO to relay radio signals around the Earth dates from Arthur Clarke's crucial article in 1945, the world had to wait until 1957 to see the first artificial satellite in orbit around the Earth. In that year, the Soviets launched Sputnik, not into a geostationary orbit, but into an orbit with apogee at 947 km and perigee at 248 km. It weighted 83.6 kg and carried two radio transmitters.

The next step was to experiment with a ready-made satellite, the Moon, and in 1959 radio signals from the radio telescope at Jodrell Bank in England were bounced off the surface of the Moon and received nearly three seconds later in Cambridge, Massachussetts, in the USA. The experiments were repeated a year later with a man-made device: Echo-1, a large balloon which reflected radio signals between New Jersey and California.

The stage was thus set for the start of the satcoms development more or less as we know them today. The first satcom capable of relaying messages immediately (or 'in real time', as the jargon has it) was launched in 1962—Telstar, which was funded by American Telegraph & Telephone Company (AT&T). It was not, however, in geostationary orbit; it orbited the Earth at heights between 950 km and 5600 km, an orbit known as highly 'elliptical'. This enabled it, for example, to connect ground stations in the USA and UK for only about half an hour each day, and it provided the first live television transmission across the Atlantic.

Before pursuing the rapid development of telecommunication over the past 25–30 years, however, it would be advisable to speak a little about their general characteristics, and which factors are important in determining their use. First and foremost is the orbit of the satellite: all the power in the world will not help if the satellite cannot 'see' the part or parts of the Earth with which it is intended to communicate. Something has already been said about orbits in general (see Chapter 1), and we are primarily concerned with GEO (where the satellite remains permanently over the same point on the earth), and the Molniya (where the satellite is in a highly elliptical orbit and its apogee, where it travels the slowest, can be so arranged that it will linger in sight of the required region). There are, of course, many variations of these two orbits, but most commercial satellites are perched side by side around the equator in geostationary orbit (Fig. 6.1).

With three satellites in geostationary orbits, it is possible to see (or to 'illuminate', as is more often said) virtually the whole of the Earth, except the polar regions, as is the case with the Inmarsat global system (see Fig. 6.2). The area of the Earth's surface visible from a geostationary satellite is a circle with a radius of 9050 km around a point on the Earth directly under the satellite. This means a circle extending from 81.3 ° North to 81.3 ° South in latitude, and between 81.3 ° East and West in longitude. The satellite is 'visible' from every point within the circle. amd we shall see later how important this is for telecommunication by satellite. Because of atmospheric attenuation, however, for satisfactory signals to be exchanged, the satellite needs to be a few degrees above the horizon.

A satellite in the Molniya orbit is complementary to those in GEO and, as Fig. 1.5 shows, can illuminate the northern or southern latitudes, or, indeed, by varying the orbit inclination, any other area of the globe. All this presupposes that the satellites' antennae are correctly positioned in relation to the Earth.

Next must be considered the power of the satellite signal. By way of establishing a reference point, it should be recalled that the first commercial satcom, Telstar, produced a signal of only two watts. Power on board a satellite is still, for obvious reasons, at a premium, but modern satellites are able to produce several hundred times more power than that. Nevertheless, the power becomes dissipated when it is directed towards the Earth with the wide ('global') beam, and use is increasingly made of antennae which give

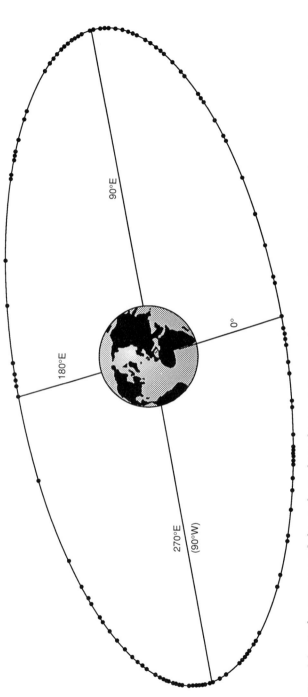

Fig. 6.1 Each point around the circle marking the geostationary orbit represents a satellite in its allocated 'slot'. Slots are in great demand and interference between neighbouring satellites is a constant problem

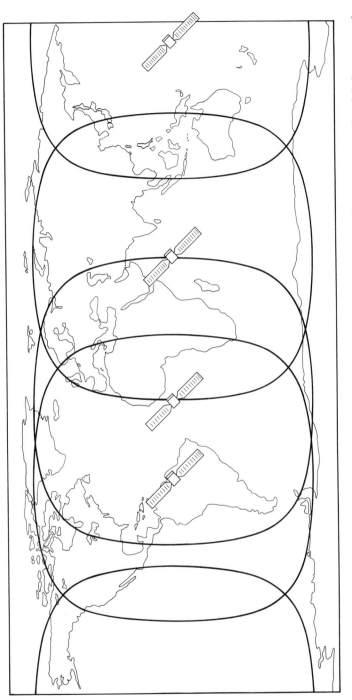

Fig. 6.2 By deploying its satellites to give four overlapping areas of cover, Inmarsat is able to over virtually global coverage for communications with ships, aircraft, and vehicles

a small 'footprint' on Earth, often shaped to put the maximum power precisely where it is needed. These are often referred to as 'spot' beams, and the technology has become extremely sophisticated (Fig. 6.3).

Clearly, the greater the power of the signal transmitted from the satellite, the smaller the antenna on the ground needed to receive it. Thus, the large ground stations built to receive signals from the global satellites providing worldwide communication are often up to 30 m in diameter. At the other end of the scale, an antenna as small as 45 cm in diameter may be adequate to receive a television signal from a direct broadcast satellite using a spot beam to concentrate its power on the target area.

And, finally, the frequency used for transmitting to and from the satellite has to be carefully selected. Each frequency band has its advantages and disadvantages, but in any case the choice is by no means left entirely to the

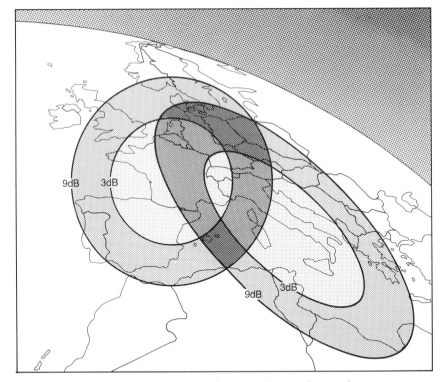

Fig. 6.3 Communication satellites, such as ESA's ECS, frequently use antennae which are tailored to concentrate the signal on the ground onto a specific target area. These are called 'spot beams' and enable the use of smaller receiving antennae on the ground

satellite operator. The atmosphere is full of signals jostling with each other to pass to their destinations, like shoppers at a Christmas sale, and allocation of frequency bands is made at regular World Administrative Radio Conferences (WARC) operating under the ITU, which, it may be recalled, is also responsible for the allocation of orbital slots in the geostationary orbit.

Lack of frequency space is still the biggest of the satcom operator's headaches. They are in practice obliged to operate in either the C or Ku bands. With satellites using the C band (and these are in the majority), the signal from the ground to the satellite has to be accommodated into a bandwidth of 500 MHz around 6 GHz. The return signal from the satellite to the ground is allowed the same bandwidth around 4 GHz. Until comparatively recently, this restricted bandwidth automatically imposed a physical limitation on the number of signals which could be carried. At that time you need 40 kHz of bandwidth to carry the human voice; therefore 80 kHz for a full telephone circuit. Thus, within the allowed 500 MHz, the satellite operator could accommodate 12 transponders (transmitters), each carrying 1000 voice channels (half a circuit), or, since television required one voice channel per line on the TV screen, one TV programme. The operator's satellite would therefore be limited to carrying 12 TV programmes or 6000 telephone conversations, or some equivalent mixture.

However, spurred on by the shortage of bandwidth, engineers have contrived to squeeze more signals into the same space. The first breakthrough was in the polarization of the radio waves.

The electrical field can be either in the horizontal or vertical plane, and two signals can in this way be sent simultaneously using the same frequency. A second improvement in signal capacity came through modification in the coding techniques in which the messages are transmitted.

Previously, the coding was done by modulating a carrier signal emitted at a particular frequency—the so-called frequency modulation (FM) method. Newer methods involve coding the signal into digital forms, i.e. impulses of 0 or 1. Each of these impulses is known as a bit—a binary digit—and eight such bits are needed to transmit a single character. The ingenuity of the engineers is such that it is now commonplace to be able to transmit 60–100 million bits per second over a single transponder; something around ten thousand printed pages. When one now notes that a single satellite can easily have fifty or more transponders, the capacity becomes staggeringly high.

The demand for capacity is such, however, that the engineers are tending more to the higher frequencies, and experiments are already being flown in the Ka (26.5–40 GHz) band and even at 60 GHz. There are, however, many imponderables by way of attenuation of the signal through rain, snow, and even clouds. These phenomena have already been experienced at the lower frequencies, and it must be reluctantly admitted that the atmosphere is by no means benign so far as satellites are concerned.

Apart from the disruptive effect of precipitation on satellite signals (the attenuation increases with the frequency of the signal), the signal to and from a geostationary satellite is interrupted whenever, as viewed from the ground station, it passes in front of the Sun. This happens twice a year, at the equinoxes, and the radio noise from the Sun smothers the satellite signal for up to six minutes on four consecutive days. At other times, disturbances may be caused during periods of increased solar activity.

But perhaps the most familiar of the drawbacks of satellite telecommunications is the delay introduced by the time the signal takes to travel from the ground to the satellite and down again to the other party on Earth. This distance being around 80 000 km, the time required is inevitably around 0.3 seconds; and the same, of course, in the reverse direction. In the early days of satellite telecommunications, this complicated conversations except between regular users, who knew to accept a certain discipline in their telephone calls, and above all not to chip in when the other party was speaking.

Once again, the engineers have come to our rescue, and on a properly adjusted circuit, it is difficult to know whether one is speaking over a satellite circuit or not.

It is usual to classify satellite telecommunication services as follows:

(1) fixed satellite services, where signals are relayed between fixed Earth stations, which are relatively large, complex and expensive (Fig. 6.4). The Earth stations are connected to the conventional terrestrial telecommunications network. The service is intended for long distance telecommunications.

(2) mobile satellite services, where the signals are relayed between a fixed Earth station and a much smaller station, fitted to a ship, aircraft, or vehicle, enabling either communication between the ship and a subscriber to a terrestrial telecommunications network, or between two ships (Fig. 6.5), aircraft, or vehicles (Fig. 6.6). Maritime communications are the most developed of the mobile services.

(3) broadcasting satellite services, where the signal is transmitted from a fixed position on the ground to the satellite and then back to individual or community receivers. This service requires the generation of high radiofrequency power from the satellite in order to make reception possible with small antennae.

It is, however, wrong to consider satcoms in isolation without taking into account the ground stations (both those which transmit the signal up to the satellite and those, of varying size, which receive a signal from the satellite) as well as the terrestrial telecommunications networks. Satcoms must be regarded as a valuable tool to enhance the performance of conventional

Fig. 6.4 The Norwegian EIK Coastal Earth Station—part of the INMARSAT mobile communications network

Fig. 6.5 Tens of thousands of ships are now equipped with such an INMARSAT terminal, connecting them to the international telecommunications network

Fig. 6.6 An increasing number of vehicles are being equipped with small INMARSAT terminals, giving them access to world-wide communications

telecommunications means, but the whole is a rather complicated system which—generally automatically—will decide when it is appropriate to use a circuit provided by a satcom.

Fixed satellite services

The fixed satellite services were the first to develop, and there was rapid recognition that the global possibilities necessitated the creation of a new kind of international organization. This led to the creation of Intelsat, an international organization with now more than one hundred members having its headquarters in Washington, dedicated to providing fixed satellite communication services all over the world. Interim arrangements for a global satcom system were agreed as early as 1964, but it took until 1973 for the so-called Definitive Agreements to be negotiated and enter into force.

Each signatory makes an investment based primarily on its use of the system and the capital has been used to purchase successive generations of satellite systems. The revenue which these satcoms generate provides running expenditure and return on the invested capital.

The first Intelsat satellite, known as Early Bird, was launched in 1965. It provided 240 telephone circuits or one television channel. Technological progress, and the phenomenal increase in traffic, can be judged from Table 6.1, showing the size, power and capacity of the seven generations of Intelsat satellites so far produced.

The success of the Intelsat services inevitably prompted the formation of regional satcoms, first for domestic traffic in the US and then spreading to Europe, Asia, and Australia. Intelsat signatories are obliged to co-ordinate with Intelsat before using non-Intelsat satellites for their international fixed services traffic. Systems have to be shown to be technically and economically compatible with the Intelsat system. In plain words, the regional systems must not interfere with Intelsat's satellites, nor must they cause Intelsat revenues to be significantly reduced.

To see the extent of the development of regional satcom systems, it is worth looking briefly at some important, and typical, examples.

Eutelsat

The European Telecommunications Satellite Organization was provisionally founded in 1977, with headquarters in Paris, by representatives of the European Conference of Postal and Telecommunications administrations, and given the task of establishing and running a European satcom system.

Table 6.1 Increasing capacity of the seven generations of Intelsat

Intelsat	1	2	3	4	4A	5	6
Year of first launch	1965	1967	1968	1971	1975	1981	1986
Transponders	2	1	2	12	20	29	46
Power (in watts)	40	75	120	400	500	1 200	2 180
Mass in orbit (kg)	39	86	152	732	863	950	1 700
Voice channels*	480	480	2 400	9 000	12 000	25 000	40 000

* In the later generations the capacity is considerably expanded by the use of techniques briefly discussed above. For example, Intelsat 7 will be capable of providing over 200 000 telephone circuits.

The constitution and financing of Eutelsat are modelled on Intelsat. Although the formal legal tests did not come into force until 1984, Eutelsat started in 1977 to cooperate with the European Space Agency (ESA) in the exploitation of their experimental satcom OTS (Orbital Test Satellite), and in the design of the operational satcoms in the ECS (European Communication Satellite) series, which ESA were planning, and which in due course were taken over and operated by Eutelsat. Eutelsat is currently procuring its second generation of satellites. Although conceived primarily for long distance telecommunication traffic, a significant part of Eutelsat's revenue has come from relaying of television programmes for distribution through the terrestrial network or into cable systems.

Arabsat

The Arab Satellite Communication Organization was founded in 1976, with its headquarters in Riyadh, Saudi Arabia, by 21 members of the Arab League. It is designed to provide telecommunications and television relay services throughout the Arabic-speaking world, and to this end it has acquired and operates its own satcom system.

Palapa

The Palapa satellites have been operated by the Indonesian authorities since 1976. Indonesia, with its 5000 islands spread over 5000 km, is ideally suited to the use of satcoms. The second generation of Palapa satellites, in addition, offers capacity to other ASEAN* countries.

Aussat

Early in 1982 Aussat Pty Limited, a company created apart from the Australian Overseas Telecommunications Commission, was awarded a contract for the first generation of the Aussat satcoms, and the first satellite was successfully launched in 1988. A second generation contract was awarded in 1988 with launch dates scheduled from 1991.

---

* ASEAN—Association of South East Asian Nations: Brunei, Indonesia, Malaysia, Philippines, Singapore, and Thailand.

The USSR

The Soviet Union was not a member of Intelsat, and in 1971 created a similar organization—Intersputnik—to provide satcom fixed services for its 14 member states and for a number of other countries where appropriate ground facilities have been installed. Intersputnik uses the various families of USSR satcoms: Molniya, Raduga, and Gorizont in much the same way as Intelsat operates, except that use is made not only of geostationary orbit, but also of highly elliptical orbits, to one special variety of which—as has been mentioned previously—the Molniya satellite has given its name.

Mobile satellite services

In the early 1970s, discussions started in the International Maritime Organization (IMO) or Intergovernmental Maritime Consultative Organization (IMCO), as it was then known, over the possibility of using satcoms to improve maritime communications, not least for safety purposes.

Towards the end of 1973 the IMCO Assembly convened an international converence to decide on the principle of establishing an international maritime satellite system, and to conclude the necessary international agreements. This international conference concluded its year and a half's work in September 1976 by the adoption of what has become the Inmarsat Convention, as well as the complementary Operating Agreement.

Both agreements, which charge Inmarsat always to act for exclusively peaceful purposes, came into force on 16 July 1979, 60 days after states representing 95 per cent of the investment shares had subscribed to them. Thus Inmarsat was born. It now has over 60 member states.

It is worth quoting from Article 3 of the Inmarsat Convention, in order to understand the scope of Inmarsat's original remit from its participating states:

... to make provision for the space segment necessary for improving maritime communications, thereby assisting in improving communications for distress and safety of life, the efficiency and management of ships, maritime public correspondence services, and radio determining capabilities.

To provide an early service to its potential users, Inmarsat took advantage of existing satellite developments by leasing Marisat satcoms from ComSat in the United States, and Marecs satcoms from ESA. There were later supplemented by maritime communications payloads flown on Intelsat satellites, to give a total of eight dedicated payloads strategically placed in the geostationary orbital slots so as to give near global coverage. The system went into operation officially on 1 February 1982.

The steady growth of traffic obliged Inmarsat to order four satellites for its second generation satcom system. The first of these satellites was launched in October 1990 and the remaining three came into service progressively during 1991. Each will be capable of handling at least 150 simultaneous telephone circuits, and this number can be more than doubled by the use of new digital techniques. Inmarsat 2 satellites have four to five times more capacity than the satellites of the first generation which were leased from ESA and ComSat.

Even so, the traffic growth is such that Inmarsat was obliged to solicit offers for its third generation of satellites even before having taken delivery of Inmarsat 2. Inmarsat 3 will be significantly larger—about ten times the capacity of an Inmarsat 2 satellite, and will have spot beams, in addition to global beams, in order to cater for the new aeronautical and land mobile services.

The relatively new possibilities of providing profitable services to vehicles has encouraged a number of private venture companies to spring up. They offer services based on the ability to provide accurate positioning information through satellites. Geostar and Qualcom are two such examples, although, sadly, Geostar and its European counterpart Locstar, have both gone into bankruptcy. The basic service is for a truck or a train to be located at regular intervals. The information can be made available either centrally, to allow fleet control, or to the individual vehicle. Several such systems have been developed: Inmarsat C terminal, ESA's Prodat, as well as others from the USA. Most systems allow a short message to be sent with the location information.

In addition to position location, satellites are increasingly being used for navigation. This application has been spurred by the US military Global Positioning System (GPS) which has so far been made available free of charge to the civil sector, though not in such an accurate mode as is used by the military. The full GPS will use over 20 satellites and there are several civilian schemes to include payloads on satcoms which can improve the reliability of the GPS system for the civilian user. Not unexpectedly, there is a comparable Soviet system, Glonass, and there have been some commendable but inconclusive efforts to work towards compatibility between these two expensive systems.

Land mobile communications are, of course, by no means all satellite-based. In many parts of the world, cellular ground-based telecommunications are serving both drivers and pedestrians. These services are now enjoying a considerable boom, with competing services helping to keep prices down. It is unrealistic to expect satcoms to compete with cellular systems; satcom land mobile communications can be useful, however, as a complement to land-based systems, either by linking cellular systems, or by providing a service remote from the larger cities and main highways, in areas where a ground-based system could not operate economically. The market will in the end decide, and operators must achieve the best mix of ground-based and satellite communications to give the cheapest and most reliable service.

Broadcasting satellite services

Because of their ability to transmit a signal to an infinite number of points within their area of visibility, satcoms are ideal for the transmission of television, either directly into individual homes—the so-called direct broadcasting—or to community antennae, which then pipe the signal by cable to the subscribers in the vicinity. The satcom regional systems described above, such as Eutelsat, have television programmes as a major part of their traffic, but the signal is, of course, transmitted only to fixed points for further distribution by conventional means, and the modest power of the satellite necessitates the use of larger antennas than are suitable for private installation.

For many years, research money on both sides of the Atlantic was spent on developing the high power travelling wave tubes (TWTs)—around 200 watts of power—believed to be necessary for transmission of wide bandwidth television signals from geosynchronous orbit. The belief died hard, and even when the Canadian Technology Satellite Program (CTS) demonstrated in 1976 that excellent quality pictures were possible with TWTs of only 20 watts, the evidence was largely ignored. High-power TWT production is still something of a black art. Programme managers who need to procure 200 watt TWTs for their satellites go to enormous lengths to life-test their TWTs before flight, and to weed out those which exhibit inexplicable eccentricities; they come to know each TWT like a shepherd knows his individual sheep. Even so, TWTs remain one of the weakest links in the chain, as French satellite operators in the late 1980s found to their cost.

Ironically, after all the money spent on research and development, there are distinct signs that the higher power is not necessary. Many believe that the future lies in 40 or 50 watt TWTs.

What is certainly clear is that the main difficulties surrounding direct television broadcast by satellite are virtually all non-technical: present day satellite technology and the advances in ground equipment design are sufficient to guarantee a high quality picture under virtually all circumstances. The difficulties arise more in the commercial and regulatory fields. Commercial systems, of course, depend on subscriptions from the viewers, as well as advertising fees. This means that the transmitted signal must be coded to avoid illicit viewing, and the subscriber is issued with a decoder. The economics of direct broadcast television are extremely volatile, and several proposed systems have failed to raise the necessary financial support. In many countries the regulatory background is being liberalized, and this will undoubtedly persuade new promoters to try their luck. In one way or another, therefore, it seems certain that more direct broadcast television stations will appear, and that the battle will be fought out between the high-powered marketing teams, rather than on the technical differences between the various satellite systems.

The development of satcoms

Although some US firms, such as AT&T, were quick to take advantage of the advent of satcoms, the main contribution to the early development of satcoms came, not from the private sector, but from space agencies. Organizations such as NASA in the USA, the European Space Agency (ESA) and the Japanese NASDA, were all obliged to devote considerable resources to the technologies of communicating reliably between the ground and their satellites, which, in the 1960s, were being launched in increasing numbers on scientific missions. It was necessary to communicate with the satellites in order to control their position, and to give instructions to the various sub-systems responsible for the satellite's health. Moreover, the satellites were acquiring more and more data.

As the design of the scientific experiments became more complicated, the scientists sought to have access to the data 'in real time', so as to interpret them quickly and send back to the satellite any necessary instructions such as, for example, new pointing directions for telescopes. The spectacular advances made in coping with vastly inflated data streams have been directly inherited by those now interested in satellite communications in its own right. The space agencies pioneered many of the techniques of antenna development and increasing satellite capacity which are the life-blood of today's satellite industry. The rate of technological change has been such that few commercial industrial firms could afford to invest in the necessary research and development without the financial support of national and regional space agencies, and this remains the case.

It is noticeable that when financial problems with other programmes forced NASA to reduce investment in satcom research and development in the period 1974–1982, European industry, which continued to have the support of the ESA, considerably improved its comparative competence in this very competitive field. In the 1990s, however, the pace will be set more by satcom operators. Space agencies will nostalgically strive to obtain funding for new generations of satcoms, but it is unlikely to be a priority area in view of the enormous private sector investment.

7. Meteorology and Earth observation

When the first cloud pictures were received from satellites nearly thirty years ago, it was quickly realized that they could reveal valuable information not available from the meteorological stations on Earth. The world's first meteorological satellite—Tiros 1—was launched by the USA in early 1960 and the prospects that it opened up led the United Nations General Assembly to ask the World Meteorological Organization (WMO) to study the progress made in research on the Earth's atmosphere and how weather forecasting might be improved, in particular through the use of satellites. Fortunately, therefore, there has been a steady tradition of international cooperation in this field, with the space agencies showing themselves very responsive to the needs of the meteorological organizations. A complete global network, known as the World Weather Watch (WWW), has been in operation since the later 1970s, thanks to contributions from the USA, ESA, Japan, and the Soviet Union.

WWW made use of two polar-orbiting satellites, in orbits about 900 km above the Earth, and five satellites in geostationary orbit. In this way, complete global coverage could be obtained. Meteorological satellites ('meteosats') are now routinely used to allow wind forces and directions to be calculated, the type and height distribution of clouds, moisture content of the atmosphere, temperature gradients and suchlike data dear to the hearts of meteorologists. Modern meteorological satellites are used not only for relaying images and other data produced by their instruments, but also to collect data from automatic meteorological stations, and to disseminate meteorological data. They therefore provide their own telecommunications capability.

The basic instrument used in meteorological satellites is the radiometer. The radiometer on board the current series of ESA meteorological satellites, for example, consists of a telescope (with a focal length of 3650 mm) with a set of detectors in its focal plane to measure the radiance of the Earth and its cloud cover in the visible, water vapour absorption, and thermal infrared spectral bands. An image in each of the three spectral bands is sent back every thirty minutes. Figure 7.1 shows a montage of three of these images; the colours are not natural ones and have been chosen to bring out the differences.

Fig. 7.1 ESA's METEOSAT satellites provide images of a large slice of the Earth every 30 minutes. In addition to the optical band—familiar through television meteorological reports—the images can show either infra-red or water vapour content

As with satcoms, it is important to bear in mind that satellites are only one part of the increasingly complicated system for providing reliable weather forecasts. They also play a significant role in climatic research, and there is increasing demand for more sophisticated instruments to be flown. It is arguably artificial to separate meteorological satellites either from scientific satellites (because of their common interest in the scientific aspects of weather and climate) or from Earth observation satellites. The justification lies in the unique organizational framework which already existed for meteorology, not only in the form of national meteorological offices, but also the WMO, to which reference has already been made.

This organizational structure has facilitated international cooperation and, in Europe at least, has led from the first experimental satellite to the relatively easy birth of a regional organization—Eumetsat—dedicated to the establishment and exploitation of European systems of operational meteorological satellites. Eumetsat, created by 16 national meteorological authorities in 1986, took over the first generation of satellites (Meteosats) developed by ESA (the first was launched in 1977), and established itself in Darmstadt in the Federal Republic of Germany. A second generation of meteorological satellites is under development, with a first successful launch in early 1991.

In the USA a similar function is now performed by the National Oceanic and Atmospheric Agency (NOAA), although the earlier US meteorological satellites were developed and launched by NASA. In the 1980s there was a half-hearted attempt to transfer responsibility for US meteosats to the private sector, but common sense prevailed and the large public service element was recognized. Transfer to the private sector would have seriously limited the contributions of the US civil meteosats to the current problems of the environment.

As most viewers of television will have seen, the development of weather fronts, or storms and hurricanes can be vividly displayed by playing consecutively the images received approximately 50 times a day from the geostationary meteosats. This is of value to the expert, as well as entertaining the lay viewer, for the information is more accurate and rapidly available than before.

Meteorological satellites have to provide an operational service: satellite data must be available without delay for blending with information received from other sources, such as coastal radar, buoys, aircraft, and even human observation. At the same time, meteorologists are as ambitious as all other scientists, and there is a constant drive to fly improved instruments, provided that they have a high probability of operating in space.

We need, before leaving meteorology, to note the progression through medium- to long-term weather forecasting to climatology, and the contribution that the whole of this area of space activity can make to the gigantic task of monitoring our environment.

Earth observation

Observing the Earth from satellites is, after telecommunications, proving itself to be the second important area of the successful application of space techniques. Instruments are now regularly carried on spacecraft to, as one says, remotely sense the Earth's atmosphere and surface. Indeed, these techniques can also be used to reveal interesting figures even below the Earth's surface.

Earth observation spacecraft are generally placed in a polar orbit 800 to 1000 kilometres above the Earth. The instruments therefore pass over or near to the poles on each orbit, and because of the Earth's rotation, they scan different sections of the Earth's surface. (Experts may wish to add that the slight flattening of the Earth around the poles also disturbs the spacecraft's orbit and that, without intervention, each orbit would be slightly different from the preceding one.) An orbit of this type takes around 100 minutes and on one side of the globe it will be travelling to the north (the 'ascending' track) and on the other towards the south (the 'descending' track). This means that the local time at which the instruments view a particular point can be varied so that the Sun's angle and intensity are the most favourable to viewing. The favourite is a Sun-synchronous orbit which ensures that the instruments view the same point on the surface at the same local time on each pass.

But what, in fact, are we looking for? Or, more exactly, what are we seeing? Every material on Earth reflects, absorbs, or emits electromagnetic radiation. The extent to which it does so is dependent on its physical characteristics. The individual radiation pattern of a material is known as its 'spectral signature'. This radiation is a form of energy which can travel through a medium, such as the atmosphere, or through a vacuum; measuring instruments in aircraft or spacecraft can therefore measure the radiation and interpret the 'signature'. The radiation received will have components of different wavelengths. For example, the visible wavelengths provide information useful to cartographers, whereas infrared radiation can show the state of health of forests and crops.

Remote sensing in its simplest form is passive, i.e. the instruments are designed just to receive and record the radiation signals from Earth. 'Active' instruments are also now being developed on the principle of radar which will bombard the target with microwaves and then record the reflected signal.

There is, of course, some scattering or absorption of the radiant energy as it passes through the various media separating the target from the measuring instrument—known as a sensor. It is thus necessary to understand the response of radiation in different parts of the spectrum, so that signals can be correctly reconstituted and interpreted.

The data accumulated by the imaging sensors are transmitted to the Earth as a series of numbers, i.e. in 'digital' form. Each image, or 'scene' as it is sometimes called, consists of a network of rectangular picture elements—'pixels'—for each of which the sensor will attribute a digital value corresponding to the average strength of the radiation the sensor has received from the corresponding rectangle on the Earth's surface. One pixel in an image received from the French satellite SPOT* (Système Probatoire pour l'Observation de la Terre), for example, covers an area on the ground of 20 m by 20 m. 20 m is therefore said to be SPOT's spatial resolution. It means, quite simply, that it is capable of distinguishing objects down to 20 m in length or breadth. Within the surface of the area contained in a pixel there could, however, be considerable variations, so that the averaged pixel reading conceals some interesting features. Hence the drive—apart from defence needs to be able to distinguish a tank from a lorry—for better resolution, and resolutions of 10 m and less are become more common even in civil satellites.

The spacecraft's sensors are, of course, mounted so that they are pointed towards Earth during flight, and they receive radiant energy from a predetermined width of ground. A satellite is said to have a 'swathe' of so many kilometres. The first ESA Earth observation satellite, ERS 1, was launched in July 1991, has a number of instruments, each with its own swath. For example, the Active Microwave Instrument (AMI) has a swath width of 100 km, and a spatial resolution of 30 m. As Fig. 7.2 shows, the swath can be situated exactly below or to the right or left of the satellite's track on Earth. In some cases, the instrument can be tilted to change or to increase the area covered.

Of course, one can approach the problem of interpretation from the ground. By identifying on the ground the spectral signature of a crop, a mineral deposit or a particular type of forest, the data received from the sensors can be sorted so as to identify all areas on the image which possess this signature.

In essence, therefore, a satellite earth observation system consists of the following elements:

- one or more space-borne sensors;
- the means of transmitting the data digitally to one or more receiving stations on Earth;
- substantial computer facilities which receive, process and enhance the data; and
- the support of ground link data and the end users of the data.

---

* It is convenient to refer to SPOT as if it were a single satellite. However, thus far, five satellites have been approved in this series.

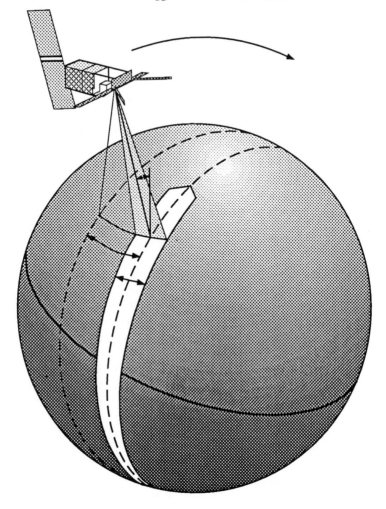

Fig. 7.2   The Earth-observing instruments are designed to view a swathe of the Earth's surface and then to move on to the adjoining swathe, until complete coverage is obtained

The choice of the spacecraft orbit and the types of sensors to be carried will depend primarily on the mission or missions. All-purpose Earth observation satellites are unfortunately not yet with us, and the spacecraft has to be chosen so as to obtain the data in the most appropriate bands and taken over the desired areas at the optimum times, and with the required repetition rates. The designers will also take into account the complementarity with the measurements from different instruments. For example, the geometric accuracy or images may be increased by flying an altimeter coupled to the imaging sensor.

It is not practical to list all the types of sensors in use or development, but the following examples will demonstrate the range of present possibilities.

1. HRV sensors—high resolution sensors operating in the visual band and the infrared part of the spectrum. This type of sensor has been the workhorse of the Earth observation satellites. The Spot satellite, for example, has two identical HRVs and they can operate either in panchromatic mode (what we might call 'black and white') giving a spatial resolution of 10 m, or in multispectral mode (where data are obtained in several frequency bands) which produces a spatial resolution of 20 m. There is the possibility of pointing the sensors 27 degrees east or west of the local vertical axis, and, by combining images taken at different angles, producing stereoscopic— apparently three dimensional—images.

2. Thematic Mapper (TM), as flown on the satellites of the US Landsat series. This sensor operates in seven frequency bands with a spatial resolution down to 30 m. The swath width is 185 km. The reader will probably have already spotted the advantage of combining the data from a SPOT HRV— which with a spatial resolution of 10 m can provide basic structural information for the image—and the very much wider spectral information available from the TM—seven spectral channels, as opposed to three. The exercise is not, however, without its difficulties, as will be seen when one considers the ensuing complications in the equipment on the ground necessary to receive and process the signals received from the satellite.

3. ATSR—the Along Track Scanning Radiometer—such as flown on ERS-1 (the ESA satellite). This is a passive instrument, that is to say, wholly dependent on radiant energy received from the Earth, using a four-channel radiometer providing measurements of sea surface and cloud-top tempera- tures. A new scanning technique enables the Earth to be viewed at different angles (0 degrees and 52 degrees) in two swaths 500 km wide and about 700 km apart. The data are then combined to give an accurate calculation of sea surface temperature, free from atmospheric distortion. There is also a two-channel microwave sounder to give information on the total water content in the atmosphere, within a 20 km footprint. This, too, is used to improve the accuracy of sea surface measurements.

4. Altimeters—for example, that flown in Topex (topographical experi- ment). This provided a precision of 20 mm, and atmospheric correction was provided by an on-board microwave altimeter. A radar altimeter is also to be flown on ERS-1, and it will be used to measure wave height, wind speed and sea surface elevation.

5. CZCS—Coastal Zone Colour Scanner—was an instrument flown on the US satellite Nimbus 7 which provided valuable data from 1978 to 1986 on the colour of the oceans. Such data are highly prized by oceanographers for the valuable information provided on a whole range of subjects, including chlorophyll distribution, suspended matter, plumes in estuaries, turbidity, algal biomass production, plankton concentration, etc.

6. SAR—Synthetic Aperture Radar. This is a technique very much in vogue, and the trend was set by a US satellite, Seasat, which was in operation in 1978 for only 100 days. The data it produced are still being avidly analysed by scientific groups in many parts of the world. (An example of SAR imaging from the ERS 1 satellite is given in Fig. 7.3.) Conventional

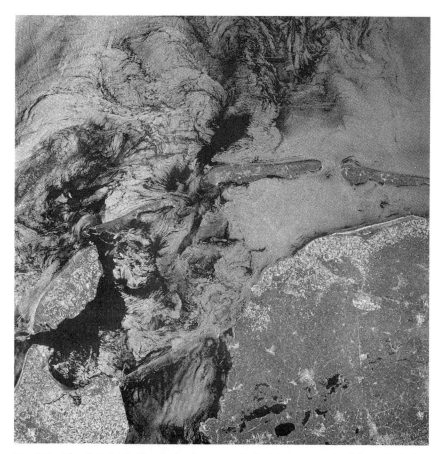

Fig. 7.3 The first SAR (Synthetic Aperture Radar) image received from ERS 1 in mid-1991—a view through cloud and darkness of part of the Netherland's coast line

radar receives the reflected signals along the length of an antenna and integrates them to produce a single image. In space, however, this would lead to impossibly large antennas—around 2 km in order to arrive at a resolution of 25 m. Instead, synthetic aperture radar for this resolution would use an antenna only 10 m long, and the signals it receives as the spacecraft travels through 2 km of orbit are integrated to give a single image.

Since the untimely demise of Seasat, Earth observation specialists have been impatiently awaiting the arrival of data from a satellite which will provide a continuous supply of radar images. Radarsat, the Canadian programme, is a brave attempt to combine federal and regional government funding with a contribution from the private sector. After many vicissitudes (the UK was for some time a potential partner, but finally withdrew), the funding now seems assured. However, in 1990—several years before the launch of Radarsat—the Soviet Union launched Almaz 1, a large radar-imaging satellite weighint 18.5 tons, and with a resolution of 15 m. The Soviet Space Agency (Glavkosmos) has announced that data from Almaz will be sold commercially worldwide. It is important to remember that such satellites will be able to acquire images in all weather and during darkness, which will make it much easier to acquire complete sets of land coverage in areas often partially covered by cloud.

The ground segment of the Earth observation system is no less complicated and almost as expensive as the space segment. The basic principle is that the data are transmitted to Earth whenever the satellite is within sight of a ground receiving station; between times the data are stored in on-board recorders, although this is not always possible when large quantities of data are involved, as for example with SAR. (It must be noted that these ground stations are different from the ground control, which, as the name indicates, is where the satellite itself is controlled and monitored, as distinct from the reception of the remotely sensed data.)

For polar-orbiting satellites, data reception facilities are favoured in the high latitudes, because they see the satellite during more orbits than those nearer the equator. The ground station at Kiruna, in northern Sweden, for example, is able to receive data from ERS-1 during 10 out of 14 daily orbits. It is usual, however, to take down data at more than one station. In these days of increasing commercialization, this inevitably means that the ground station must pay for the data, or, as is often the case, provide some service which its geographical position makes possible.

The receipt of data on the ground is the beginning rather than the end of the story. After acquisition the data have to be put into a form suitable for archiving. Much more data are acquired than can be completely processed; ioreover, the degree and type of processing depend on the needs of the nd-user. For example, the ESA centre at Frascati, Italy, where Earth observation is archived and turned into products for users, holds over 90 000 images just from the Thematic Mapper—one of the instruments mentioned above.

The data are stored on tapes or, increasingly, on optical discs, and retrieved whenever a user has need of them. According to requirements, the user can request Computer Compatible Tapes (CCTs) suitable for viewing and further manipulation on the user's own computer, or single images which may be composed from data received during several orbits.

Computer capacity alone is not the answer; considerable time, expense, and ingenuity is involved in producing software—the operating instructions for the computers which will transform the data into a form which is meaningful to the end user. Specialists produce algorithms (formulae) which convert streams of seemingly meaningless digital information into two- or even three-dimensional representations.

Most common of the techniques is to allocate colour codes to values, so that all pixels showing the same value are represented in the same colour. It is sometimes tempting to believe that the beautiful images we so often see are in natural colours, but they are 'false colours', simply used to facilitate a visual understanding of the underlying data.

The applications of remote sensing satellite data are many and increasing rapidly, but they can mostly be slotted into one or more of the following three categories:

• survey, mapping and inventory activities;

• monitoring dynamic changes on land and sea; and

• change detection and disaster surveying.

But increasingly there is a call for a more systematic collection of data. This involves asking a series of questions about the types of information that are needed: the accuracy, the spatial resolution, the frequency, and many other factors. The kind of sensor to be used and the orbit in which it is flown considerably influence the sort of data that can be gathered. It is not as simple as taking a single 'photograph' from space from which all users can derive whatever information they may need; would that it were so. In fact, it is necessary for end-users to specify their requirements and ambitions and for these to be transformed through a series of discussions with spacecraft engineers and instrument designers. Only in this way is it possible to formulate instrument and satellite specifications in the knowledge that the data obtained will truly be of help to the multitudinous authorities.

Data from Earth observation satellites are a contribution to the solution of important terrestrial problems—pollution, crop disease, illicit urban development, desertification, and so on—but it is vital that the space data be compatible with data acquired by other means. Nowhere is this so true as in the area of cartography, where a great deal of work is necessary to remove the inherent distortions in satellite images, so that the satellite data can be directly converted into an accurate map where none existed before, or perhaps laid over an existing and outdated one.

The present move is towards geo-based (or geographical) information systems (GISs) containing data of various types and from different space-borne sensors, as well as other sources. The end-user should be able to ask for seemingly unrelated sets of data to be pulled out of the system and superimposed, in order to see whether any previously unsuspected relationship exists.

The potential benefits for mankind are obvious, and Earth observation satellites are destined to play an important part in the new drive to clean up the environment. But, in spite of the sensational advances in sensor and software development, there is still insufficient coordination. The lack is not in the technology or the science, but in the sheer organizational and political difficulties in bringing this capability to bear on the problems.

8. Military uses of space

Not to deal, however briefly, with the military uses of space would leave the reader with an incomplete and distorted picture of the overall situation. In most countries, the defence budget paid for the original investment which converted bomb-carrying missiles into launch vehicles capable of putting spacecraft into orbit. Without such massive injections of cash, the civil development of space would have been significantly delayed; space scientists would have been obliged to continue their research using ground-based techniques or, at best, balloons and high-flying aircraft.

But the justifications for including a military chapter are not just historical; throughout the past forty years military requirements have forced technology ahead in virtually every space-related discipline. Perhaps military ambitions were no more exacting than those of the space scientists, but the defence budgets were many orders of magnitude greater than those for civilian science.

Military space activities can be divided into six categories: communications; meteorology; surveillance and reconnaissance; navigation; geodesy; and finally the rather separate category of space-borne weapons and defensive mechanisms. Before looking briefly at each of these categories, it might be useful to have some impression of the size of military space activities.

Without burdening the reader with specific dollar amounts, it is well to bear in mind that the US defence/space budget started to overtake the NASA budget some years ago, and the same trend seems to prevail in the USSR.

Dr Bhupendra Jasani* reports that, between 1958 and 1987, 2499 'military-oriented' satellites were launched by the USA, USSR, PRC, France, and the UK.

Military satellites have always greatly outnumbered civil satellites. During the period 1967–87 there was a fairly steady average of 100 military satellites launched each year, from just the USA and the USSR. Eighty or ninety per cent of these launches were from the USSR, and this disparity can to some extent be explained by the Soviet policy of using large numbers of satellites in low orbits, which limited their life. This characteristic reflects the

---

* Jasani, B. (1988). *Space and international security*. Royal United Services Institute for Defence Studies.

greater availability of launches in the Soviet Union, and perhaps also the fact that the Soviet aerospace industry has not yet been able to achieve the same long-life reliability which has been developed in North America and Europe.

It is easy to see that such a volume of satellite and launcher production has given to the US aerospace industry a solid production basis from which civilian satellite manufacturers have been quick to profit. That they have not similarly profited in the development of civilian launch vehicles was due, as has been explained elsewhere, to the US policy of discouraging civil development for fear of damaging the intended commercialization of the Space Shuttle.

Before looking at the different categories of military satellites, it must be emphasized that a trend has developed towards multi-use satellites and there are many examples of basically civil satellites either giving a service to military users, or even carrying a separate military payload. Conversely, several military satellite systems are available, in whole or in part, for civilian use, until pre-empted in the event of military need.

Communications

This category needs little explanation; defence needs for reliable and safe communications are obvious. In the USA there are several specialized defence satcom systems: Fleet Satellite Communications (FleetSatCom); Air Force Satellite Communications (AirSatCom); Defence Satellite Communication Systems (DSCS), etc., and it is safe to assume that the Soviets have a not dissimilar inventory.

These communication satellites have orbits and use frequencies suitable for their specialized needs, whether this be for regular peacetime voice and data transmissions, or for tactical satellites launched to provide communications in an inflamed area. NATO has long had its own system of satcoms, as has the UK, and France is similarly developing a military system after having used a defence payload riding piggyback on a civilian satcom system.

Obviously, as with civilian satcoms, there needs to be a corresponding ground network, with the added problems that the military planner has to solve in addition to those of his civilian counterpart, not least the ability to operate in spite of jamming, as well as to frustrate unfriendly attempts to read the signal. At any one time there are at least 60 military satcoms in operation.

Meteorology

Here again, the military have needs beyond those of their civilian opposite numbers, and this has given rise to separate military meteorological satellite systems in both the US and the former USSR, although clearly there is a good deal of commonality. Defence interests are not only in real-time weather information, as well as longer-term predictions, but also in improving the accuracy of long-range missiles, which is affected by meteorological activity along the trajectory. The US Defense Meteorological Support Program (DMSP) makes use of satellites in near-polar, Sun-synchronous orbits at a height of 850 km, from which each satellite orbits the Earth 14 times daily, viewing a swath of about 3000 km.

As is the case with satcoms, considerable use is made of data received from civilian meteorological satellites, but the military planners always have the additional worry of ensuring that the data on which they are dependent will be available under all foreseeable circumstances—a guarantee civilian satellite systems are unable to give.

Surveillance and reconnaissance

Spacecraft are used to provide both strategic and tactical information for military commanders at different levels in the command structure. The orbits are chosen so as to maximize the results from the sensors carried on the spacecraft. It is convenient to divide the missions into four self-explanatory categories: photographic; electronic; surveillance of the oceans; and early warning.

The same techniques are used as in civilian Earth observation satellites but, once again, there are additional military requirements. Whether or not it is possible to read a vehicle registration number in Red Square, the military sensors have undoubtedly achieved a very much higher resolution than is available—or perhaps even needed—in civilian spacecraft, but apart from this increased sensitivity, the military require rapid availability of data in a usable form. Their spacecraft also need to be protected both against natural hazard and, to the extent possible, from man-made interference and attack. No wonder that the cost of the latest so-called 'spy in the sky' satellites greatly exceeds that of conventional spacecraft.

The importance attached to being able to survey the oceans, day and night and in all weathers, has given a tremendous push to the development of space-borne radar, which can see through cloud cover and darkness. There is no doubt that the development work paid for by military budgets and the defence interest in having access to the products of civilian satellite systems have greatly accelerated the availability of radar aboard civilian satellites.

Early warning satellites are deployed by both the USA and the former USSR. They generally carry a selection of sensors, among which there is invariably one or more infrared devices capable of detecting very small temperature changes (less than $1°$ C) on the ground. These are used to detect the launch of missiles, transmitting the information back in real time, either directly to a ground station, or through a link to another communications satellite and thence to the ground.

In one way or another, therefore, satellite technology has made it extremely difficult to concentrate troops, prepare an attack or even to build defence-related factories or installations, without almost instant detection and recognition by others. This capability is, of course, very relevant to the international disarmament and verification negotiations, and there is growing pressure for Europe to develop its own, independent satellite system able to provide European governments with their own sources of information.

9. Risks and legal aspects

Space debris

Space activities since the launch of Sputnik I in 1957 have led to 3600 artificial satellites being put into orbit by 3000 launches. However, in addition to these intentional satellites, a large number of unintentional satellites—from used launcher stages to ejected lens caps—has been produced. Because of the potential danger to other satellites (and because they could be mistaken for aggressive missiles), the US Space Command—and formerly the North American Aerospace Defense Command (NORAD)—keeps track of them all. A total of around 20 000 such objects have been tracked, and of these 7200 are today still in orbit; only 5 per cent are operational satellites. The total weight of all this space junk is estimated at two million kg.

Just how virulent this proliferation can be is to be seen by the unexplained explosion of an Ariane third stage in 1986; this alone produced 470 pieces of debris (Fig. 9.1). Using powerful ground based radars, objects as small as 100 mm or less can be tracked on a regular basis. Additionally, objects down to 10 mm can be tracked on a non-operational basis, using optical telescopes.

Given that these objects are travelling at anything between 5 and 15 km/s, even the smallest particle has the capability to cause catastrophic damage to a spacecraft. With the present estimate of smaller, untracked objects currently at between 30 000 and 70 000, it is understandable that space agencies—not to mention insurance companies—are extremely concerned about the risk of collision. Both the US Space Shuttle and the Soviet Salyut 7 have already had windows damaged by space debris, and there are growing numbers of stories about 'near misses', such as the crew of the Space Shuttle Columbia which, on a flight in 1982, watched a burned out upper stage of a Soviet rocket as it passed by only 12 km away, at a speed of 11 200 km/h.

Interest in space debris is increasing because of the growing number of manned spacecraft either already in orbit or planned for the near future. It has been estimated that the probability of a large space station having a collision with a piece of space debris is about 2.1 per cent for a mission of ten years.

Optical detection from outside the atmosphere should provide the best data

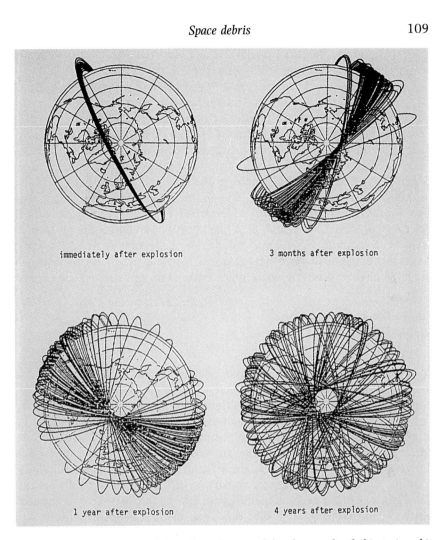

immediately after explosion   3 months after explosion

1 year after explosion   4 years after explosion

Fig. 9.1 Reconstructions of the orbits of some of the thousands of objects in orbit around the Earth, and which have to be avoided by man-launched satellites. These were caused by the destruction of a malfunctioning rocket.

on space debris, and the Jet Propulsion Laboratory has suggested putting small telescopes on board a spacecraft which would look down on the near-Earth debris, sending data back to Earth where computers could calculate the size and orbit of pieces.

In addition to the risk of collision, astronomers are concerned at the growing disturbance that space debris causes to their observations. Spurious trails are frequently found on images of the sky taken through ground telescopes and the glint from satellites is thought to be a likely cause of many non-meteoritic flashes detected by observers on the ground.

Space debris is now widely accepted as an international problem requiring international solutions, but there has so far been no agreement on specific legislation, although space agencies cooperate regularly both on data exchange and on techniques and good practices which would at least slow down the rate at which debris is increasing. Ideas are beginning to develop for space debris garbage collection. One such scheme would attach a drag device to the debris to greatly increase the drag and thereby cause its orbit to decay more rapidly. The cost of such operations, however, means that they would have to concentrate on the smaller numbers of relatively large objects. This would, of course, be a help, but there remains the problem of what to do about the much larger number of smaller objects which present a continuing hazard to manned and unmanned spacecraft alike. It seems as though we have already taken our bad habits with us on our journey into space.

Space insurance

In the early days of space activities, neither launcher nor satellite was insured. The programmes were all either scientific, developmental or, at most, pre-operational, and the funding authorities were content to live with the possibility of a launcher failure or, indeed of the satellite itself not functioning as planned. Launcher manufacturers contracted with their clients to 'use their best efforts' to ensure that their rockets worked successfully, and launch failutes were accepted stoically, since there was no alternative. Similarly, scientific satellites were built by companies which could theoretically be sued for negligence in the event of malfunction, but the satellites were not insured on the open market against that risk. Even today it is not usual to insure against the loss of a scientific satellite, principally because they are generally single spacecraft designed for a particular function or event, and the insurance money would not enable a timely replacement to be procured, however consoling it might otherwise be.

The move towards insurance—towards sharing the risk of failure—started with the advent of commercial satcoms. The satellite owners were responsible to their shareholders in the same way as any other commercial enterprise, and if the satellite failed to become operational in orbit, revenues were lost, and not 'just' scientific data. Hence the recourse to the insurance market, not only against failure of the launcher, but also total or partial failure of the satellite's payload in orbit, including a shortened life-time. (Insurance for injury or damage to third parties caused either by launcher or satellite is mandatory under a United Nations convention.)

The challenge of this new business was enthusiastically accepted by the insurance world, and in most places, it was allocated to the aviation insurance specialists. They were soon to learn that the analogy was not an exact one. In the early days the brokers simply did not have access to the expertise necessary to assess the risks involved either with the launcher or the satellite. The first insurance policies were rather simple, premiums were low and compensation was paid promptly on the minimum of evidence.

The end of the 1960s and the beginning of the 1970s, however, produced failure rates over 30 per cent, and this changed the insurance situation completely. Brokers started to specialize in space insurance and to engage space experts as consultants or staff members. This in turn led to a considerable increase in the demand for information on the launch vehicle and satellite, and the way in which they were being produced and handled.

Policies became much more complicated and new insurance formulae were devised in many ways paralleling the 'no claims bonus' and 'third party liability' familiar to car drivers. But the early 1980s produced another crop of launch failures and of satellites which, although successfully launched in LEO, failed to make GEO or just did not function in orbit. New in-orbit failures early in 1991 are likely to force up the cost of in-orbit insurance from 2 per cent of insured value to 2.5 per cent. This would mean an increase of many hundreds of thousands of dollars per year for communications satellite operators.

Premium rates had already risen as a result of the early failures, but the second wave resulted in extremely high rates—exceeding 30 per cent at the peak in 1986—in an attempt to cover previous losses. Cumulative deficits, that is to say, losses in excess of premiums, built up to US$240 million, and there was an understandable reluctance of the market to take on space business above a very limited amount. At one period the world market was reluctant to have more than US$200 million at risk at any one time, but this has subsequently increased to around US$300 million—enough to cover a launcher and two fairly large satellites—and is still rising slowly. The trouble is that the financial capacity for covering space risks has to come from other sectors of the market, where earnings are much higher, and where the extent of the risk is perhaps easier to assess, and certainly much easier to explain to potential investors.

In spite of setbacks, the insurance market still exists and premiums are now down below the 20% mark; there are about 25 companies which undertake space insurance, although two-thirds of the business is cornered by a handful of regular groupings. Two of the insured satellites which failed in orbit were recovered by the Shuttle and—now the property of the brokers who paid out on the losses—after refurbishment, sold to and re-launched by new clients. This was certainly an encouragement to the insurance market, and most of them believe that space insurance will gradually develop into just another line of conventional insurance.

Current launch insurance rates are between 16 per cent and 20 per cent of the insured value, but both brokers and customers are ingenious in devising new insurance options. One such option, known as 'buy in the sky', allows satellite operators to defer payment for satellite manufacture and launch until the satellite is functioning in orbit. The insurers have now become established members of the space community and their continued presence is essential to the extension of commercial space activities.

Legal aspects of space

Jurists define space law as regulating relations between states, international organizations, and private persons arising from the exploration and exploitation of outer space. Space law had its origins forty or so years before the first artificial satellite was launched. A Belgian lawyer at the Court of Appeal in Brussels published an article* on the subject in 1910 in which he propounded the need for a specific legal system in readiness for the time when scientific and technological progress made space flight possible.

Two decades later, Vladimir Mandl, a lawyer from Pilsen, published a monograph† in which he reviewed the legal problems connected with space flight. Although it stimulated no interest among his contemporaries, Mandl's work was surprisingly far-sighted. He foresaw, for example, that launchings would need to be carefully regulated, and that spacecraft would need to have a nationality. He and Korovin, who wrote in the same vein a couple of years later, were the pioneers of space law, but it was necessary to wait another two decades before the subject was seen to be of practical importance, and therefore worthy of systematic study. Throughout the 1950s and the early 1960s, the legal implications of space activities attracted a growing number of lawyers, mainly those already specializing in air law.

Those who were occupied in building rockets and satellites or developing scientific experiments had neither time nor interest for the learned arguments as to whether or not the principles of Roman law were more applicable to space activities than the Chicago Convention on International Civil Aviation. However, much of the theorizing was eliminated at the end of 1966, when the General Assembly of the United Nations approved the Outer Space Treaty: the first of a series of important international treaties attempting, as Laude had predicted half a century previously, to provide an international legal system to regulate space activities.

---

* Laude, E. (1910). Comment appliquera le droit qui regira la vie dans l'air. *Revue juridique internationale de la locomotion aerienne*, No. 16.
† Mandl, V. (1932). *Das Weltraumrecht: Ein Problem der Raumfahrt*.

Credit for this work goes to the United Nations Committee on Peaceful Uses of Outer Space (COPUOS) which, together with its technical and legal sub-committees, painstakingly strove to develop legal texts during a period when the main space powers were scarcely on speaking terms.

The other fruits of the work of COPUOS are:

(1) Treaty on the Rescue and Return of Astronauts and the Return of Objects launched into Outer Space, 1968;

(2) Convention on the Liability for Damage caused by Space Objects, 1972;

(3) Convention on the Regulation of Objects launched into Outer Space, 1974; and

(4) Agreement governing the Activities of States on the Moon and other Celestial Bodies, 1979.

Obviously, this set of UN treaties and conventions alone is not sufficient to cope with the growing number of areas in which space activities are a necessary component, but it deals with the particular characteristics for which no existing body of international law was appropriate, and it allows other specialist areas of law—telecommunications, patent, etc.—to be applied where appropriate.

Specialized legislation, and indeed, national legislation, thus has a basis on which to build when endeavouring to adapt to the new circumstances produced by the application of space techniques.

The notion that outer space and celestial bodies are free for exploration and use by all states, and not subject to national appropriation, may today sound banal, but its universal acceptance represents a remarkable victory for common sense over national bigotry.

The principles accepted by the UN work have left a lasting stamp both on space law and its derivatives, such as the conventions establishing the international satellite telecommunications organizations Intelsat (for fixed communications by satellite) and Inmarsat (for mobile communications by satellite), which have both espoused the UN principle that 'communications by means of satellites should be available to the nations of the world . . . on a global and non-discriminatory basis'.

The International Telecommunications Union (ITU) has provided the regulations necessary for telecommunications by satellite, through inclusion in its International Telecommunication Convention and Radio Regulations of a mechanism for resolving conflicts of interest. The procedures apply to all telecommunications administrations and are administered by the International Frequency Registration Board, a permanent organ of the ITU.

All satcom operators bow with respect to the ITU's World Administrative Radio Conferences (WARC), referred to in Chapter 6 above. WARC's are convened every four or so years to grapple with the problem of frequency allocation for the growing number of different services offered through satellites. The resulting agreements have the legal status of treaties between the 160 members of the ITU.

The ITU also has committees which fix standards for the use of radio frequencies and promulgate call signs and emergency codes.

The work of the UN in this field is by no means completed. Apart from a number of fascinating theoretical problems—where does outer space begin?—the advent of international space stations brings a whole family of new legal problems. Astronauts of many nationalities may be working in laboratories belonging to several different nations; it does not need much imagination to realize that situations can arise which would pose legal problems. This sort of problem should be relatively easy to solve, but then come the somewhat more complex problems of intellectual property rights, patents, and the like.

Even operations involving 'old-fashioned' satellites are haunted by the fear that a patent infringement lawsuit brought by a US aerospace firm will involve the payment of billions of dollars for the unlicensed use of patented technology. 113 satellites are involved in this case, of which 12 or 13 are non-US. The case has, in fact, been brought against the US Federal Government, and it remains to be seen how far other governments will be obliged to make good any payments the courts decide to impose.

The role of government in such matters is a direct consequence of the UN principles which are based on an early realization that, however remote the chances, the private sector could never alone assume responsibility for compensation in case of a catastrophic accident. For this reason, governments are obliged to assume responsibility for the consequences to third parties of any launching taking place on their territory.

The production of acceptable legal principles to cover the field of remote sensing has so far eluded COPUOS. The Legal Sub-Committee, as long ago as 1975, produced a report assembling a number of points on which delegations seemed able to agree, but it has not yet proved possible to translate these into an acceptable legal text. The sensitivities are obvious: many states insist that remote sensing of their natural resources requires their prior authorization, whereas others (particularly those who have the necessary technical means) advocate the principle of 'open skies'. This is likely to remain a formally unresolved area for some time.

Similar problems await the introduction of programmes for transmitting energy to Earth from space, but fortunately we have a decade or so before this problem needs to move up the agenda.

Not surprisingly, space law has blossomed over these past 30 years, at international, regional, and national levels. Virtually every space programme now needs the advice of lawyers, not simply to ensure a satisfactory procurement contract, but also to ensure compliance with the complex net of legislation which has developed. A new breed of space lawyer has evolved, familiar with the growing body of international law and also at home with the jargon dear to the hearts of the space community. No organization involved, even indirectly, in space activities can afford to leave home without one—and generally several.

Lawyers first started to be fascinated by the new legal problems posed by space flight long before that capability had been acquired, but many problems of space law are now on the critical path to the successful exploitation of our newly acquired facility to operate way beyond the Earth and its surrounding atmosphere.

10. National and regional space activities: a brief survey

Government funded programmes

It is notoriously difficult to assemble reliable and comparable figures of governmental space funding, not only because some countries do not divulge their spending or quote it in hard currencies to equate with the major, freely exchangeable currencies, but also because their internal organization and working methods are so different from those in the majority of countries. Even in the more accessible countries it is difficult to be sure that all contributions have been included. Provided the reader is not looking for strict accuracy, however, the following table will give an idea of the total size of government space spending and of the relative annual expenditures of the major players.

Approximate civil space expenditure, 1990, in US$ millions	
NASA (USA)	12 000
Japan	907
ESA	1 730
France	1 355
Germany	801
Italy	615
Netherlands	98
Norway	31
Canada	258
India	440

Another way of judging relative space expenditure is as a percentage of Gross National Product. Figures published in 1990 showed:

USA	0.595 per cent
France	0.135 per cent
Belgium	0.052 per cent
Germany	0.050 per cent
Italy	0.054 per cent
Japan	0.040 per cent
UK	0.030 per cent

Here again, however, the figures are quoted only as an illustration and in full knowledge of their frailty. They do, however, help to show that government space expenditure, even in the USA, is not a major devourer of the taxpayer's money.

Yet another popular yardstick—particularly when trying to solicit reluctant Treasury support for more funding—is the governmental space expenditure per head of population. In 1989 it was estimated that the USA ($120.00) and the USSR ($110.00) spent virtually the same, if one includes both civil and military expenditure, whereas Europe and Japan trailed at $8.60 and $7.10 per head, respectively (source: IBF Spectrum, Vienna—August 1990).

But national and regional approaches to space activities are very varied and the following pages will be devoted to a very short description of the developments in individual countries, as well as in the European Space Agency (ESA). We start with the two 'super space powers': USA and USSR.

USA

The US civil space programme is mostly, but not entirely, funded and controlled through the National Aeronautics and Space Administration (NASA), which was created in 1958 from the National Advisory Council for Aeronautics (NACA), founded some 40 years earlier—in 1915, to be exact. Its annual budget is fought through Congress each year in a mammoth series of hearings, cuts, and restitutions, which fray the nerves of senior NASA officials and those of their foreign partners in cooperative ventures. Its many past successes and the industrial interest with its large programmes awake have protected NASA from mortal, or even seriously wounding attacks. Indeed, until the last four or five years NASA was considered to be the golden agency, and managed to defend its annual budget remarkably successfully.

It is easy to attribute the changed atmosphere to the Challenger Shuttle accident in 1986, but in fact it was probably most attributable to the NASA policy of discouraging the further development of expendable launch vehicles in order to funnel traffic on to the Shuttle. When the Shuttle had to be taken out of service for 32 months following the Challenger accident, US payload manufacturers were left without a US alternative. NASA's image was not reburnished by the several public enquiries that were held, nor has it been helped by subsequent hitches, both with Shuttle launches and, for example, the Hubble Space Telescope. People with a space background are generally indulgent in such circumstances; there is an uncomfortable feeling of 'There, but for the grace of God . . .'. Customers tend to be more hard-nosed about it; and so are other government departments who perhaps felt that NASA had been unduly protected from criticism in the past.

NASA's activities have, for many years, been coordinated with other government departments, through an inter-departmental committee of senior officials. Although this still continues to function effectively, mounting pressure induced the White House to institute a National Space Council, chaired by the Vice-President. In parallel with this, agencies such as the Department of Transportation and Commerce have started to play much greater roles in determining national policy.

As if this were not enough, NASA has undergone a significant internal reorganization and its methods of running programmes were at the end of 1990 scrutinized by the National Academy of Public Administration—perhaps as a result of the failure of the Hubble Space Telescope to perform anywhere near its exacting specifications.

But it would be wrong to believe that the demise of NASA is near. On the contrary, with a staff of approximately 20 000 it remains a strong and competent space agency with enviable experience and capability—plus a budget of US$ 12.3 billion in 1990 (up from US$ 10.7 billion in 1989).

Against this background, a high-level panel was appointed in August 1990 to review the future US civil space programme and reported to the NASA Administrator at the end of 1990; this was the so-called 'Augustine panel'. The report was widely acclaimed as a model of clarity and common sense. NASA, too, should be pleased for, although the report contained some courteously phrased criticisms, it left the space agency intact and gave no succour to the many agencies and governments which had been hoping for a dismemberment. Nevertheless, the recommendations represent a major change in direction for NASA; a reinforcement of the science and basic technology programmes, a de-scoping of the Space Station programme, and the replacement of the manned Shuttle by a new heavy launch vehicle whenever 'human involvement' is not needed.

The outside pressure is certainly so strong that, whatever resistance to some of the recommendations may exist within NASA, implementation is virtually certain; and NASA will emerge stronger for this check-up. Some organisational changes have already been made as a result of the Augustine report.

It must be admitted, however, that the shape of NASA's future programmes, and indeed its future organization, is not yet certain. Space Station Freedom has been reduced in size and stretched out in time; a new heavy lift launcher is to be started, together with the Department of Defense, but it is unlikely that any further Shuttles will be added. The distinct change of direction is illustrated by approval of a large (30 billion US dollars) Earth Observing System (EOS). Although there are continued arguments about the preferred size of the satellites in the programme, there is general agreement that higher priority must be given to understanding the Earth's environment.

Meanwhile the US President has already enthusiastically called for a lunar/Mars programme. The debate is proceeding at many different levels, with scientific and industrial pressure groups lobbying actively. The need for Congress approval of NASA's annual budget certainly does not make for stability.

Often overshadowed by the NASA programme, mention must be made of the programme funded by the National Oceanic and Atmospheric Administration (NOAA). The development of the NOAA programme depended heavily on the space technology developed through NASA, and indeed started using protoflight models of NASA's Synchronous Meteorological Satellites (SMS) for early data collection. Since then NOAA has developed a wide range of its own increasingly sophisticated satellites. The agency's satellite and associated expenditure is probably around 350 million US dollars annually, and it can be expected to play an increasing role in climate and environmental space programmes in the future.

US defence spending on space continued to increase during the 1980s, and has now overtaken the NASA budget. Within the limits imposed by security, there has always been a reasonable flow of information between the US defence and civil space programmes. This was encouraged by the usefulness to the defence community of many of the civilian satellite programmes. The Department of Defense, however, has never been enthusiastic about sharing the development costs of the Shuttle or of the Space Station; although the defence community will almost certainly want to use the Space Station facilities when they are ready, just as they now buy Shuttle flights to launch defence payloads without having contributed to development costs.

The connection between civil and defence space programmes is—paradoxically, perhaps, in view of the security barriers—closer at the level of industry. This is not to imply that security is breached; it is simply that the size of the defence programmes and the manpower they need automatically upgrade the competence of the industrial companies. Transfer of staff from defence projects provides regular injections into civilian programmes of experienced staff from the front line of technological development. US individual advantages in civil space competition with European space industry stem largely from the enormous national space programme.

USSR

For the first decade of Soviet space activities, there were no official figures about even civil space expenditure. Announcements on future programmes were rare and laconic; first news of launching generally came from a unit run by a schoolmaster with his schoolboys at Kettering in the United Kingdom. The civil space world for years used his observations as a basis for judging the type of Soviet satellite which has recently been launched and its probably objectives.

First tentative changes came with the creation of Glavkosmos, usually described as the Soviet Space Agency, although the space sciences, life sciences, and the applications satellites and experiments were produced by independent institutes under the Academy of Sciences.

More recently, however, there has been considerably more openness, and organizations and committees which apparently shape Soviet space policy are beginning to identify themselves. Senior engineers, who never appeared in public, and whose names were only occasionally to be found in the literature, are attending international space conferences and taking part in inter-governmental negotiations.

Defence and civil space are still linked more organically than in the West, but there is considerably improved visibility for the foreign observer. More-over, the same debate as in the West over manned and unmanned space, and the impact of the current economic and political climates, were regularly ventilated in the Soviet press and in statements by Soviet spokesmen. The combined effects of the harsh economic situation and the dramatic lessening of East–West tension probably produced a reduction of 20 per cent in Soviet space spending in 1991, but the overall Soviet satellite launch rate remained stable at 75 in 1990, compared to 74 in 1989.

With the disintegration of the monolithic Soviet Union as we have known it for more than 70 years, it is difficult to foresee whether a Soviet Union space programme can long be maintained as such. Many key laboratories, factories, installations, and launch sites are located in the independent republics. But space in the USSR has always been closely linked to defence, and it may therefore—at least in the immediate future—be maintained as a central function.

Russia is offering services in several different areas: training and flight of foreign cosmonauts, launch services, flight of microgravity experiments and remotely sensed data from their Earth observation spacecraft, as well as technological developments. This is seen as one of the former Soviet Union's few sources of hard currency revenue in the field of high technology, though it remains to be seen how the space programme survives the political disintegration of the Soviet Union.

All this seems to indicate that, without significantly reducing their space-related infrastructure either in orbit or on the ground, the former Soviet Union is backing away from space spectaculars, and planning to concentrate more on the utilization, and even the sale, of space techniques. The emphasis is likely to be on research and development, and on programmes which are directly relevant to practical problems. It probably heralds, too, a move to suggest more international cooperation in developing such things as future space stations or planetary exploration programmes. This may well lead to a resurrection of the Soviet Union's earlier proposal for a World Space Agency, although the USA and most Western European countries are by no means favourable to the idea. In their view, international collaboration is best achieved through specific space programmes, rather than by creating another international organization. Once again we see emerging the two hard-to-reconcile themes of international cooperation and international competition. This time the competition will not be in terms of impressing the public with Moon landings and the like, but in a fight for space-related business. But more of this later.

European Space Agency (ESA)

Still some way behind the two space giants, Europe is widely acknowledged as the third space force, particularly if one takes together the programme of the European Space Agency and the national programmes in its member states. Founded in 1975 out of the fusion of two space agencies formed in the early 1960s,* ESA now has 13 full member states: Austria, Belgium, Denmark, France, Germany, Ireland, Italy, Netherlands, Norway, Spain, Sweden, Switzerland and the United Kingdom. In addition, Finland is an associate member and Canada is connected to the Agency through a special agreement.

ESA's programme is divided into two categories: the Mandatory Programme, to which all member states are obliged to contribute according to their Gross National Product; and the optional programmes, each of which is the subject of a separate agreement signed by all member states who wish to participate, and specifying the scale of contributions agreed between the participants.

The Mandatory Programme covers the Agency's general budget which includes such items as future studies, general technological research, fellowships, and the scientific programme. It should not be forgotten that ESA's progenitor ESRO started life exclusively for scientific projects—first using sounding rockets and then graduating to increasingly large satellites.

---

* European Launcher Development Organization (ELDO) and European Space Research Organization (ESRO).

The current ESA scientific programme is known as Horizon 2000, and is the first consistent attempt to produce a scientific programme which uses the limited annual funding to meet the priority needs of all the various branches of space science.

The following table shows the approximate percentages with which the 13 members states contributed to the ESA mandatory budget in 1990.

Austria	2.31
Belgium	3.05
Denmark	1.94
France	18.15
Germany	23.25
Ireland	0.50
Italy	13.91
Netherlands	4.91
Norway	1.97
Spain	6.22
Sweden	3.50
Switzerland	3.98
United Kingdom	16.31

The optional part of the ESA programme consists of all the other Agency activities—from the Ariane programme, through Hermes and Columbus, back to Earth observation and telecommunications programmes. Although this system is undoubtedly very heavy and involves a considerable number of rather complicated legal documents, it has the significant advantage of binding participants to finance a programme right through to the end. An inflation formula is built into each agreement and a participant can only withdraw if the cost to completion will exceed 120 per cent of the original commitment—excluding inflation. (A much more satisfactory way of financing long development programmes than NASA can boast, with its annual battle with Congress over not only the total NASA budget but even the proposed allocations to individual projects.)

An important feature of the ESA system is the principle of 'fair return'. That is to say the placing of industrial work in member states in proportion to the financial contributions to the programme. For many years, member states were more or less content to accept a minimum return of 80 per cent of their theoretical entitlement. However, as the competence of the industries in the smaller member states improved, and their appetite for more sophisticated work increased, the Director-General was obliged to take special measures aimed at bringing all members up to, or within sight of, 100 per cent.

It is easy to point out that this geographical distribution of industrial work can increase the cost of a programme, but it should not be overlooked that it has always been one of ESA's (and ESRO's before 1975) prime tasks to raise the competence of the European aerospace industry *and* to reduce differences in standards between member states. There have been many adjustments to the rules governing 'fair return' and more may be in the pipeline, but it seems unlikely that the basic objective of levelling out industrial competence over member states can ever be abandoned—even though in some of its forms it may be held to be contrary to the European Community Single Market Act, due to enter into force at the end of 1992.

The 1990 budget of ESA is around 2 billion 'Accounting Units' (AUs). The AU is a unit specific to the Agency,* and for 1990 was equivalent to approximately £0.674 sterling, making the 1990 budget around £1.348 billion, roughly one sixth of the NASA budget.

The Agency currently has approximately 2000 staff members, distributed between the headquarters in Paris and major establishments in the Netherlands, Germany, and Italy.

The Medium-Term Plan for the period 1991–1995 to be considered by the Council of representatives of Member States, requests annual budgets which—inflation apart—show a sharp rise to cover the cost of the major programmes:

Year	1991	1992	1993	1994	1995
Million AUs	2288	2828	3098	3240	3212

The European Ministerial Council was expected in November 1991 on the basis of a long-term plan prepared by the Director General and his staff, but Ministers decided to take another year before deciding on the expensive Columbus (space station) and Hermes (space plane) programmes. They did, however, give their approval for a significant Earth observation programme. These decisions confirm two tendencies for the 1990s: disinclination to commit high proportions of space budget to infrastructure, rather than application, programmes, and a general tightening of the purse strings.

As with NASA, this debate over the ESA future programme does not in any way foreshadow the decline of the Agency. It is generally admitted to have performed well and it is a case of slightly altering course and not of fundamentally putting into question the need for the European Space Agency.

* The AU is converted into national currencies at the rate prevailing for the European currency unit (Ecu) in June of the preceding year.

Most member states of ESA have some form of national space activity. The proportion of the national budget spent on ESA programmes compared to the national space programmes varies from country to country, and over the past twenty years there have been significant fluctuations even within a single country, reflecting the priority given to space and the space organizational structure.

Against this background, a few comments will be made about the space activities of the 13 member states.

Austria

Although closely associated with the activities of both ESRO and ESA, it was only in 1987 that Austria was able to become a full member of ESA. Its contribution to the ESA Mandatory Programme is, according to GNP, over 2 per cent, but Austria's contribution to the optional programmes is generally more modest—usually around 0.5 per cent. At this level it is difficult to find packages of attractive high technology work for Austrian aerospace industry, the development of which was undoubtedly handicapped by Austria's late arrival as a full member of ESA.

Nevertheless, it is hoped that by allocating more money to Austria's national technological support programme and by invoking the help of the ESA machinery, it will be possible to improve Austria's industrial return and thereby increase Austrian government satisfaction at the results of the decision to join the Agency as a full member.

The Austrian Space Agency has a 1990 budget of 20 million US dollars, some 90 per cent of which is used on ESA contributions.

Belgium

From the very early days of ESRO, Belgium has played a significant role in European Space. In the early years this was primarily through the enthusiasm and commitment of a series of Belgian ministers who held together the European space community in a series of ministerial conferences culminating in the Bruxelles Ministerial Conference in 1975 which led to the creation of the ESA.

But Belgium's interest began to take a practical turn, both in the form of an industry capable of undertaking sophisticated work and through a government commitment to the ESA optional programme at a rate on average twice that of Belgium's GNP contribution. Belgium has thus become a contender for the title of the 'biggest of the smallest' among the ESA family. Its 1990 space budget was US$ 113 million—up 14 per cent from 1989.

There are recent signs, however, that Belgium's enthusiasm for the costlier space infrastructure programmes is cooling, under pressure from rival budgetary priorities.

Denmark

Denmark has been a loyal and low-profile member of both ESRO and ESA from the start. Its industry has acquired a sound position in one or two good technological niches, but its contribution to the ESA optional programme is always well below the GNP percentage, which is just short of 2 per cent. Scientists and industry benefit from modest government support, but there is no significant Danish national space programme, although scientific cooperation with the USSR has increased in recent years.

France

By far Europe's most active and committed space nation, France has had a national space agency, CNES,* since 1962. Autonomy in space has long been regarded as a key to maintaining a position as a major power. This aim does not entail complete French independence in all the various space-related activities, since this would be a gigantic financial burden for a single European country, but it has always been the underlying motive for France's constant pressure for the member states to agree to ESA filling critical gaps and making Europe progressively less dependent on the USA in space.

This policy has been maintained in spite of changes of government, and funding has increased steadily with, at worst, periods of stability. The CNES budget in 1991 shows a 13 per cent increase over 1990, in spite of other competitors for government funding. It follows, therefore, that France has the largest European national space programme, as well as being the largest contributor to the ESA optional programme.

The national programme, including satcoms and direct broadcast television satellites and the Spot series of satellites, consumes around 60 per cent of the total French expenditure on space. There is, in addition, a strong manned element which developed out of the long-standing scientific cooperation with the USSR.

Critics often refer to the 'Franco–French' European space programme, and it is undeniable that the general direction of ESA activities has been determined more by France than by any other nation or combination of nations. It is also true that the French industry and space agency have

* Centre National d'Etudes Spatiales.

profited handsomely from ESA funding. It is, however, no less true that without the insistence and innovation of the French, the ESA programme would never have amounted to much. They have often been unsympathetic and selfish partners; but they wanted to succeed in space and they chose to do it in large part through European cooperation. If they were sometimes misguided in the direction they bullied and cajoled other member states of ESA into taking, no one else had the government interest and support to push through an alternative programme.

Germany

Over the years, the Federal Republic has been significantly influenced by France in its space policy. This influence, born of the post-war reconciliation between the two countries, has not always been easy to reconcile with a desire to cultivate closer relations with the United States. The resulting dichotomy has led to programmatic compromises which have attempted to accommodate French 'go-it-alone' aspirations for Europe with those—led by the United Kingdom and Germany—who urged collaboration with the United States. It is perhaps possible to discern a reassertion of a more independent German space policy brought about by the political and financial consequences of German reunification.

Until 1989 Germany had no space agency. Policy was defined by the Federal Ministry of Research and Technology using the German Air and Space Research Establishment (formerly known as the DFVLR and now mercifully reduced to DLR) as its executive arm. Space science was undertaken in a number of important institutes around the country, many of which were under the mantle of the Max Planck Institute.

Founded in 1989 as a limited company under German law, DARA,* the Germany Space Agency, had a difficult birth. Neither the responsible Ministry, nor the DLR, was entirely happy with the creation of the new agency, for both lost staff and areas of competence.

It was too early to say that the move has been successful, but with the increasing complexity of space programmes, it seems sensible to have a space agency and to give it sufficient authority to be able both to formulate a national space policy and to look after its interests in the Council and other delegate bodies of ESA.

Germany's space expenditure in 1990 amounted to around DM800 million, and its commitment to the ESA long-term programme will undoubtedly cause this to increase over the next five years. More than 70 per cent of the expenditure goes to ESA and Germany is currently the second biggest

* DARA—Deutsche Agentur für Raumfahrtangelegenheiten.

contributor to ESA. The remainder has been spent on a more restricted national programme than in France: the main areas of expenditure being telecommunications and direct broadcasting, science (Germany has funded, either alone or with the USA, several ambitious scientific satellite programmes), microgravity (using Spacelab and sounding rockets), and preparing—more consistently than any other ESA member—for the advent of the Space Station. Recent indications are that Germany will aim to reduce to 60 per cent the amount of its budget spent on ESA programmes and that national remote sensing and environmental protection activities will be increased. However, such a switch is dependent upon a significant increase in the overall budgetary allocation, and the cost of German reunification—joyous event though it undoubtedly is—will not help.

Ireland

Ireland is ESA's smallest contributor, but its accession to the ESA convention in 1975 was part of the wider drive to develop closer links with continental Europe. Both in science and in the newer technologies, Ireland has been able to carve out interest areas of activity within the ESA programme. As is often noticeable in other European spheres, the Irish field competent delegates who make a conscious effort to further the European cause, and they thoroughly deserve any benefit their membership brings to them.

Italy

For more than 20 years the Italians wavered—depending on the degree of political support for space—between claiming to be the 'smallest of the biggest' or the 'biggest of the smallest' in the ESA community. They lay in fourth position in the contribution table. During the last few years, however, the UK has not significantly increased its contribution to ESA, and Italy has sailed by to become third, with a share of about 18 per cent—significantly more than its GNP share.

The organization of Italy's space activities was for long a mystery to most foreigners and to many Italians. It was largely dominated by a small number of strong personalities who managed in one way or another to find funding at the right time to enable Italy to join all the main ESA programmes. But for years, Italy's industrial return was unsatisfactory and Italy could not really be called a happy ESA member.

In parallel with the ESA membership, Italy managed to sustain an interesting national programme which still included a strong scientific element, a national satellite programme and a launch site off the coast of

Kenya. 1990 saw the launch by Ariane of Italsat, Italy's first national communications satellite with on-board switching capability, enabling it to deal with 12 000 telephone circuits, as an in-orbit telephone exchange.

A novelty, with which Italian space scientists have been associated by more than twenty years, is the tethered satellite: a satellite which is suspended from the Shuttle orbiter, or later from the space station, at the end of a long cable. In this way the altitude of the scientific experiment can be varied and the resulting measurements compared. The first of these tethered satellites has been developed in collaboration with NASA, and is due to fly in the Shuttle in late 1991. The tether can be reeled out to distances of up to 125 km, enabling the satellite to achieve low Earth orbit, which the Shuttle itself cannot reach.

Much—but by no means all—of the mystery in Italian space affairs has been dispersed by a series of reorganizations in the Italian aerospace industry, and by Italy's following the fashion of forming its own space agency.

The Italian Space Agency (ISA) came into being in 1988, and is responsible for national and international space programmes, which are funded through a series of Five Year Space Plans. In 1990 the ISA budget is 822 billion lire, which converts to approximately £370 million.

Italy has made a significant financial and technological contribution to the European space programme, but has not so far managed to develop an influence proportional to its size. It may be that the new space agency is the instrument which has been lacking. Certainly, Italy has a lot to offer, including an industry which has improved beyond recognition in the last two decades, and one can expect to hear more from them in the future; they will not be prepared to serve on as—as they sometimes see it—second class members of ESA.

The Netherlands

The Netherlands is the home of ESA's largest establishment—ESTEC—the Agency's technical centre which now houses more than a thousand staff as well as a more or less permanent body of contractors. With this (literally) concrete example of the fruits of European space cooperation on its doorstep, the Netherlands has managed to be a regular, if unadventurous, supporter of the ESA programmes.

Dutch space policy is nowadays developed and administered by the Netherlands Agency for Aerospace Programmes. The 1990 space budget was just short of 100 million US dollars, practically all of which was spent on contributions to ESA programmes.

The Netherlands has a remarkable record in astronomy, and whatever funding is available for national space programmes is invested in scientific satellite projects in collaboration with foreign institutes.

Netherlands aerospace industry is principally concentrated around the longstanding Fokker Aircraft Company, which has developed several 'space' specialities (notably the fabrication of large solar panels to provide power for spacecraft), and the space expertise does not seem to have been diffusing significantly into other companies.

On the other hand, however, Netherlands is home to one of Europe's most advanced remote sensing institutes, which has done much to develop techniques and provide training for the developing countries.

In a typically quiet way, therefore, the Netherlands have pulled their weight in the European space effort, and have built up an enviable competence in those areas which have either fitted with their, so to say, pre-space abilities, or which would be useful in furthering their aims in other fields.

Norway

Norway became a full member of ESA together with Austria in 1987, after having followed the ESRO/ESA scientific programme since its start. Like Austria, it now has its own space agency in the shape of the Norwegian Space Centre (NSC), and its 1990 budget amounted to US$30 million. 60 per cent of the space budget is invested through ESA, but Norway also spends around 20 per cent on national programmes with a further 20 per cent on cooperation outside ESA.

The NSC owns and operates the Andoya Rocket Range in northern Norway, and is in the process of making further considerable investment there in the hope of creating a launch site for the smaller launch vehicles which some analysts predict will soon start to take a larger share of the market.

Rather better placed industrially than Austria, Norway still has some difficulty in coaxing the prime contractors to give sufficiently 'noble' work to the Norwegian companies. But the national will to succeed is there, and the present steady rate of improvement will probably continue. Not unnaturally, Norway's interest is in programmes which have some scientific or technological relevance to its geographical position, its long coastline and the national importance of the North Sea oil industry. It has far less interest in space programmes of a less practical nature.

Industry is beginning to respond to the challenge (not least in the manufacture of satellite Earth and ship stations) and, provided they continue to select their projects in harmony with their talents and capabilities, there is every reason to believe that Norway will consolidate its position.

Spain

With a 1990 space budget of US$117 million, Spain has come a long way since it first became a member of ESRO, nearly thirty years ago, with a nearly non-existent space industrial capacity.

During the 1980s the Spanish government chose a novel organizational approach to space. Taking it away from the air force generals who had created Spain's space effort, responsibility was given to the Centre for the Development of Industrial Technology (CDTI) in the Ministry of Industry. The CDTI is also responsible for non-space technological development, as for example in the technology programmes (Esprit, Brite, etc.) of the European Community. The approach to ESA programmes is therefore almost entirely in terms of the technological interest and Spain's ability to win industrial work which is consistent with Spain's wider technological aims.

Spain has had virtually no national space programme in the sense that one exists, say in Germany, or even in Norway, but the Ministry of Communications has commissioned communication satellites (known as Hispasat) to provide additional telephone and TV communications, hopefully for the 1992 Olympic Games in Barcelona. It is interesting to note that the competence of Spanish industry has increased to the point where 30 per cent of the work on Hispasat is to be undertaken in Spain. The severely practical approach to space appears to be paying off.

Sweden

The Swedish National Space Board, which includes representatives from industry as well as from government, oversees Sweden's space activities. The programme is implemented through the Swedish Space Corporation (SSC) which is constituted like a private sector company.

The 1990 space budget stood at US$68 million, around three-quarters of which was committed to ESA programmes. The remainder is spent on a national programme which concentrates mainly on scientific projects, including small satellites.

The SSC has shown itself to be highly innovative and has been very active in a number of areas where commercial space activities seemed promising. In this way it has expanded the rocket launch site at Kiruna (inside the Arctic Circle) into a data reception station for satellites in a polar orbit. Its earlier efforts to find a good commercial use for the Nordic Tele-X satellite were not so successful; the difficulties of cooperating in this field proved to be intractable—even among Scandinavian neighbours. But the satellite functioned well!

The Swedish aerospace industry has a wide competence which has been steadily improved over the years, but it would appear that the government sees no reason to favour it through a substantial increase in its ESA contribution. For all its faithful cooperation with France in this field over the years (like Belgium, Sweden is a participant in the Spot programme), there is no sign that French enthusiasm has convinced the Swedish government to make space one of its priorities.

Switzerland

Always willing to pay its share and help push the European space programme, Swiss interest appears to have waned over recent years. It has a certain industrial competence and many excellent scientific institutes, but the lack of any space agency gives the impression of a fragmented effort.

This may only be a passing phase which is perhaps nurtured by the current uncertainties over the future ESA programme. There are still pockets of enthusiasm, but the overall impression is of caution.

United Kingdom

For a quarter of a century the UK shied away from the notion of a national space agency. The policy was to regard space as an instrument which each government department was free to use as it saw fit. Such coordination as was deemed necessary was achieved through a loose mechanism of inter-departmental consultation, with occasional recourse to ministers whenever money was necessary.

In this way the UK developed a three-pronged space programme: space science, satcoms, and launch vehicles, each promoted by the ministry or agency most interested. A vigorous space programme was funded by the Science Research Council (later to become the Science and Engineering Research Council (SERC). The space scientists in the 1960s were not only of world status (which many are even today), but they also had sufficient influence with the authorities to win support for the creation of ESA and a reasonable participation by the UK, and to find funding for a series of national scientific satellites.

In a separate compartment, promoted mainly by the Ministry of Defence, the UK maintained a successful missile-cum-launcher programme, which enabled the development of a successful series of launchers leading to Blue Streak—the UK's contribution to Europe's first cooperative attempt to develop a satellite launch vehicle.

The third leg—satcoms—was funded by the Department of Trade and

Industry, which understandably regards it as a promising field for commercial development.

But, lacking an underlying national space policy, these programmes proved extremely vulnerable. In the early 1970s the space scientists, always under fire from the remainder of the scientific community, had their wings clipped. The national space scientific programme was reduced, and the ESRO/ESA space science programme was capped at a level which has continued to irk the space science community for the past twenty years.

The launcher programme fell victim to government changes, aggravated by the regrettable failure of the first ELDO project to show that Europe was capable of cooperating on such a difficult, high technology venture (there was also the feeling in some parts of the UK establishment that the UK, and Europe as a whole, risked very little by relying on the US for the provision of satellite launch facilities).

The telecommunications activity fared the best of the three, for, although the specific national satcom project was subsumed by ESA, UK industry retained a significant share of the ESA satcom programmes which resulted.

With the start of ESA in 1975, the UK retained a significant role in European space activities, but with about three-quarters of the available funding devoted to the ESA programmes and a very much reduced national programme. Slowly the idea began to gain support in the UK that the increasing range of space activities and the growing relevance of space technology to non-space activities justified the creation of a national space agency.

However, it was not until late 1985 that the British National Space Centre (BNSC) was born. It was intended to regroup all the interested government departments and to produce a national space plan for the coming ten or fifteen years. Physically, the BNSC got off to a good start and, with admirable British improvization, staff were seconded from the various ministries and agencies. Sadly, by the time the plan had been produced, through an interesting cooperation between public and private sectors, the government will to proceed had evaporated. By the end of 1987 the BNSC was reduced to a coordinating agency dependent, as before, on ministerial approval for each funding venture. Earth observation now appears to be the only favoured space activity; launchers and manned space are—until the next change of policy—not considered worthy of investment.

The UK space budget for 1990 was 150 million pounds, announced as an increase of 6 per cent over 1989 (but without the reminder that inflation was running at around 7.7 per cent). The UK's total contribution to the ESA programmes dropped from 14 per cent to 8 per cent—against a GNP which had risen to over 16 per cent.

The vacillations of the UK space policy, with its mixed signals to industry and to the UK's European partners, will no doubt be the subject of a separate case study in due course. This is not the place to do more than note that the UK has played an important role in many fields of European space development. Since 1987 it has opted to reduce its contribution by about 50 per cent and its influence in ESA has suffered correspondingly.

Finland

Although an associate member only since 1987, and not participating in all the ESA programmes, it is appropriate here to make a brief reference to Finnish space activities. The Finnish Space Committee, an interdepartmental committee founded in 1985 under the aegis of the Ministry of Communications, is responsible for developing space activities in Finland.

As in Spain, the Technology Development Centre plays a significant role and, without totally forgetting a long-standing interest in certain aspects of space science and meteorology, the main Finnish interest is in the applications programmes.

The Finnish industry has proved itself well able to take up industrial work from ESA contracts, but the present level of contribution is probably too low to have the desired effect. As is so often the case, the disappointing results might persuade government to decrease their stake, even though the smallness of the stake is probably the main reason for the disappointment. The 1990 space budget was round FMk64 million.

It is to be hoped that Finland will soon join the European space community, for it has a solid contribution to make, but ESA's present concentration on expensive in-orbit infrastructure programmes is not attractive to newcomers.

Canada

Before moving to other countries it would be appropriate—though it may seem geographically paradoxical—to speak first of Canada, because in 1978 Canada signed a cooperation agreement with ESA which had the effect of making Canada a member in everything but name. The agreement has recently been renewed and extended.

Canada has had a successful space activity since the 1960s, mainly concentrated in satcoms. Like the UK, it was for many years a matter of policy not to have a national space agency. From 1969 coordination was sought through an interdepartmental committee on space. As with the UK, if a national master plan existed, it was not visible to outsiders.

After some years of discussion, however, the Canadian Space Agency (CSA) was established in 1989, and despite internal wranglings over siting the headquarters (Montreal was eventually chosen), it is beginning to make its presence felt both nationally and internationally.

Apart from its participation in some of the ESA programmes, Canada has a healthy national space programme. The two main programmes are Radarsat, an Earth observation satellite using synthetic aperture radar, and the Mobile Servicing System (MSS) which Canada is contributing to the international Space Station Programme.

Radarsat is a particularly interesting programme, not only because of its technical sophistication, but also organizationally; the programme involves Federal and State governments together with the private sector and is a serious attempt to put Earth resources satellites on at least a semi-commercial basis.

The MSS is expected to cost about 1 billion US dollars over the next 16–17 years. It is the logical extension of the manipulating arm which Canada successfully developed for the Space Shuttle programme.

The CSA's budget for the Canadian fiscal year ending April 1991 is approximately 258 million US dollars and is expected to peak at 260 million US dollars in a year or so. Very nearly half of this amount is committed to the Canadian contribution to the Space Station Freedom.

Canada's interest in space has always been severely practical and this has no doubt helped to sharpen the wits of its space industry which has a splendid record in winning export contracts. If the new CSA manages to retain government support for the present programme, Canada, over the next decade could move further up the space league table from its present position at around number eight.

Other countries

Not to add at least a few paragraphs about other national space programmes would give the false impression that space activities were distributed exclusively between North America, the USSR, and Europe. This was largely true up to the end of the 1970s, but the last decade has seen the arrival of a number of important new space programmes, the influence of which is already flowing beyond their national boundaries of origin.

China

The Chinese are fond of repeating the claim to have invented the solid propellant rocket, but China's modern space programme started in 1956 with a development programme to integrate rocket and jet engine technologies.

Since that date China has launched more that 30 satellites, including recoverable payload capsules and geostationary satellites.

The Chinese programme, being based on the military programme, has always put a strong emphasis on the development of a national launcher capability. The launcher family 'Long March' has grown steadily since its first successful satellite launch in 1970, and in its current versions is actively offering launches to Western satellite operators at attractive prices. Various new configurations of the Long March launcher appear to be under development—the visibility of the Chinese programme has been somewhat patchy since the political disturbances of 1989—aimed at putting several tons of payload into a low Earth orbit.

China now possesses four operational launch sites, together with a capable network of ground and sea tracking and control stations.

Although there are doubtless many competent Chinese space science groups, the emphasis is on the production of meteorological, remote sensing, and communications satellites. There seems, however, to be a growing interest in becoming involved in planetary missions (particularly Mars), and there is perhaps a longer term intention to develop manned space flight.

The steady progress made over the past twenty years qualifies China for inclusion in the list of space powers. Their relative isolation has resulted in the development of a virtually autonomous capability, and it is one which is itching to enter the commercial market.

Japan

The same persistence and continuity which characterize the Chinese space programme are also the hallmarks of space activities in Japan, together with increasingly large injections of cash. The Japanese space agency (NASDA) is more than twenty years old and during that time they have tenaciously followed clearly established—and announced—goals. In the 1970s the Japanese were content to rely on US technology, both to develop a launcher capability as well as to manufacture the first generation of Japanese meteorological and telecommunication satellites. During the last ten years, however, the Japanese industry was itself able to develop an impressive amount of launcher and satellite hardware and software.

The result is that Japan now has an extremely strong and diverse space programme, covering most, if not all, of the areas being developed by the USA and Europe. Not only are they developing their own meteorological, remote sensing, and telecommunication satellites, but the Japanese launcher industry is also well advanced with its H2 launcher, which will give Japan virtual parity with other major space powers. In addition, Japan is to provide a sophisticated manned laboratory module to be attached to NASA's Space

Station Freedom. As if this were not a sufficiently ambitious programme, there is also the development of Hope, an unmanned space plane which is to be launched atop H2, by analogy with Hermes and Ariane 5.

Traditionally, space science has always been separately managed—not by NASDA but by the Institute of Space and Astronautical Science (ISAS), which has developed not only its own scientific satellites, but also its own family of launch vehicles. ISAS sees itself as remaining a small and efficient scientific operation dedicated to producing the maximum scientific value out of its limited budget. For this reason it will probably be content to see NASDA make a bid to enter planetary exploration, and funds are already being sought for a two-ton Mars-orbiting spacecraft.

The way Japan has marshalled its resources and intelligently combined technology import and national development could be a lesson to many other countries. If the investment and perseverance are maintained—and there are signs that this will happen—the next years will certainly make Japan into one of the leading space powers.

India

Hard currency shortages and an abundance of well-educated engineers and scientists have combined to push India's space programme single-mindedly down the path of developing the maximum possible at home. From the creation of the modest sounding rocket facility near Trivandrum in 1963, the Indian space programme has made impressive steps, due not least to a series of gifted and devoted space leaders.

The importance that India attaches to space can be seen by the fact that the Department of Space comes directly under the Prime Minister's Office. This privileged position has enabled the Indian Space Research Organization (ISRO) to weather most of the budgetary storms and to maintain a fairly constant level of effort.

The programme was based from the start on two principal tenets:

(1) Indian space activities must directly help to solve the country's problems and thus win the support of other government departments, such as education, agriculture, and telecommunications (some people would include defence in this list); and
(2) the maximum effort must be made to use materials developed at home and to limit hard currency purchases.

The result has been a widely based space programme which has had considerably practical successes.

Here again the production of an independent launcher capability had high priority, and by 1980 India had been able to launch its first satellite. The aim is to be able to launch the second generation of domestic telecommunications satellites into geostationary orbit by the 1990s.

Several other countries have space programmes, and others are beginning to show signs of wishing to join the club.

By way of example, one might look at Israel. Space activities commenced there in 1965 with the creation of the Institute of Space Studies at the University of Tel Aviv. By 1983 the Israeli Space Agency had been founded with an annual budget of about US$4 million; but of course the Agency was able to benefit from the scientific research conducted in the universities as well as the relevant work in the industry and defence communities. The results of these efforts were first revealed to the world by the launch of two small experimental satellites in the second half of 1988, thus demonstrating a capacity both to construct satellites and to launch them.

Brazil is another example. The Brazilian Aerospace Institute (INPE) has, with French cooperation, constructed very modern ground installations for the integration and testing of satellites. A low-key collaboration with China also exists in this field, and it is probable that 1991 will see the first launch of Brazilian meteorological and Earth observation satellites.

This list is long, and will become longer; Indonesia has a flourishing space programme, and Taiwan and South Korea have been making preparations to join the club. Australia, too, must not be forgotten.

In the 1960s the rocket launch site at Woomera in the Australian desert was extensively used for European civil and military rocket development, but Australia declined to continue the association by joining ESA. The Cape York launch site (see Chapter 2) has caused a revival in interest in space, but there is little government funding available, and it must succeed as a private venture or not at all.

Whether the motive is civil, military, or both, the membership of the space club is bound to increase in the coming decade. It is also to be hoped that the newcomers take good note of the many mistakes made by the older club members, and concentrate on making different ones of their own.

11. Commercialization of space

The term 'space commercialization' is frequently used and seldom defined. In this present context, it should be taken to mean any activity directly or indirectly involving space techniques which leads to someone making a profit. Excluded are the contracts awarded to industry by civil and defence space agencies for the construction and launch of satellites for the agency's own use. This is an important exception: the US government's space contracts account for about 90 per cent of all space sales in the United States. Nevertheless, the US Department of Commerce projected US commercial space revenues in 1990 at around 3.5 million US dollars, excluding NASA and the Department of Defense.

Some areas are more truly commercial than others. Take for example the international satellite telecommunications organizations, Intelsat and Inmarsat. Both pay industry to construct and launch communication satellites, and they make revenue by providing circuits to their members— telecommunications entities in the various member states. Inmarsat, for example, is obliged to pay 14 per cent per annum in interest to its signatories and it does so—and more. This is by any standards a commercial operation.

The same is true of other, private, communication satellite organizations who, to a greater or lesser extent, make money either through hiring out their circuits or by using them to provide their own, television and other, services. In all, one can count over two hundred such commercially-operated satellites now in orbit and all analysts agree that the population will continue to increase.

This is perhaps the clearest end of the spectrum. No doubt these activities owe something to the research and development of space agencies (Inmarsat, for instance, started by leasing satellites built by Comsat for the United States and by the European Space Agency), but nowadays Inmarsat's and Intelsat's financial operations are free-standing and there is no government subsidy. The private satcom companies are, by definition, destined to make a profit, to merge with more successful competitors or to go under.

It must not be overlooked that a great deal of the business is generated on the ground. Obviously, there must be ground stations to track the satellites in orbit and to receive their signals. Less obviously there is a need for small antennae and terminals, increasingly for satellite mobile communications with ships, trucks, and aircraft. Similarly, the direct broadcast television satellites generate a lucrative, though highly competitive, market for small and environmentally acceptable house antennae. The combination of space and ground segments makes the satcom areas by far the most advanced space business.

At or near the other end of the spectrum are the companies which make a profit from improving the usefulness of images derived from space which they have acquired at a price by no means taking account of cost of developing and launching the satellite. This is not intended as a criticism: these added-value companies, as they are called, perform a very useful service to the community. Without them the use of space products would be even slower than it presently is. These cases are cited simply to point out that outside the telecoms field, there are at present very few truly commercial operations.

An honourable exception must be made of a small group of companies who offer services—that is to say, those who enable or help others to send their payloads into space. According to need, they are able to advise how to package an experiment for launch, or on a grander scale, to provide the facilities where a whole satellite can receive its last check and be prepared for flight on a launcher waiting on a nearby launch pad. The best of these companies, mainly in the USA, have survived for more than a decade—which no doubt means that their investors are not unhappy with the financial results.

The special case of launchers needs more explanation. In the United States, the expendable launch industry has largely been kept going by the regular demands of the military, who have found only a limited use for the Space Shuttle. It is difficult to speak about commercial operations in the Soviet Union, but there is no doubt that the military utilization of launchers has been the main reason for their availability for non-military missions.

Europe has not had by any means the same size of defence space programme, and the build-up of the civil launcher industry has therefore been much more difficult. It would not have been possible without the ESA having paid for the majority of the initial research and development work—even for the qualification launches of the Ariane launcher. Since 1979 a private company, Arianespace, has been successfully selling launchers and launch services on a commercial basis, but the company will not be obliged to repay the enormous investments which governments have made, not only in the launcher but also in the launch base in Kourou, French Guiana. There is some modest repayment plan when profits start to accumulate, but no one expects this to be more than symbolic.

Companies are beginning to be able to offer small launch vehicles at very much lower prices than those from the larger 'stables'. Theoretically, it is attractive for a customer to be the sole passenger aboard a rocket, but there has been some doubt whether the smaller satellite market would grow sufficiently to sustain the smaller launcher companies. Their prospects would, of course, be greatly improved if navigation or other telecommunications system were to materialize using large numbers of small satellites, instead of a few very large ones.

Remote sensing has long been hailed as the next area in which there would be a breakthrough, that is to say, where the private sector could expect soon to be making a profit. Experience both in Europe and the USA has unfortunately not confirmed that earlier optimism. In the USA, Congress insisted that the successful series of NASA Landsat Earth observation satellites should be privatized. They intended to do the same for the meteorological satellites, but mercifully this was vetoed when full realization came of the danger of leaving such a vital public service to the vagaries of private enterprise.

This private US remote sensing venture has not had an easy ride, and each year in Congress it has been necessary to fight for a continued influx of government money before the private company could afford to maintain the launch of satellites as the earlier generation reached the end of its useful life. It does not have the characteristics of a sound conventional commercial operation. In Europe, the French created a private company, Spotimage, to market the products of their Spot series of satellites. This organization has done some excellent work in making the Spot images more useful to the end-users and in the field of training; sales have increased too. However, the company is only now beginning to show a positive return on a yearly basis, and is a long way from starting to repay the capital costs of the highly expensive satellite system and its ground segment infrastructure. Spotimage has set a deadline of 1998 for becoming independent of all future government subsidies.

Some companies, however, are starting to make a reasonable living out of enhancing the images from other people's satellites and selling them to carefully targeted customers. In general the demand for remotely sensed data is growing, but not at such a pace as to promise anything near to a fully commercial operation comparable to that in the telecommunications field.

Many are hopeful that the increased interest in environmental monitoring will bring a significant increase in the use of data acquired by Earth observation satellites, and this is no doubt true. It is questionable, however, whether the environmental authorities will be willing to invest sufficient money in these satellite systems to enable anything like a commercial service to be established. This is not at all to decry the use of Earth observation satellites—quite the contrary. There is absolutely no doubt they can make a significant contribution to solving some of the pressing environmental and climatological problems. But much of this will have to be considered as public service, qualifying for some form of public funding, rather than appropriate for private enterprise. Above all, a distinction must be made between spacecraft which provide a valuable contribution to the long-term data sets needed by the scientists in their continual search for trends in such vital subjects as the state of the Earth's ozone layer, and those satellites which will give data on which action is needed at once, as in the case of oil spills or accidental releases of radioactivity. There are limits to what can be done by a single spacecraft.

Finally, a brief reminder of the prospects for business in microgravity. It would be foolish to reject this sector altogether as commercially unviable, but it would be equally misleading to give the impression that there are solid prospects for quick profits. At least in the next ten, maybe even twenty years, the benefits are more likely to arrive through basic scientific research which increases our understanding of basic physical processes which have so far resisted our research efforts on the ground. Whether this process will eventually lead to large-scale manufacturing in space remains very doubtful.

When we integrate the situation in the various potential sectors of commercialization, therefore, it is clear that, with the notable exception of telecommunications, the space specialists have promised too much too quickly.

This admission by no means closes the matter; it is simply an indication that it is essential for future predictions to be much more realistic. The disappointment which over-optimism has generated among government funding agencies must not be repeated, for fear of discouraging them altogether from further investments.

Some governments are already taking the hard line that if, indeed, there is some money to be made out of space, then the private sector should make the necessary investment. In its way, this attiude is as shortsighted as the exaggerated promises of the 1970s. The time-scale for the recovery of capital investment and the appearance of real profits is generally much longer in space-related programmes that in most classical investment areas. Furthermore, the nature of space programmes and the risk involved are unfamiliar to most investors, who have not been encouraged by recent mishaps. This is not an indirect plea for exclusively public funding for space programmes, but rather an explanation of why a rapid conversion to wholly private sector funding would not work. As has been seen above, the business prospects vary between sectors, but—with the possible exception of satellite telecommunications and its associated services—the future lies in the development of an appropriate mix of public and private funding. This will require a change in attitude by the major space agencies who have for twenty-five or more years played the leading role in space programmes, and who are used to dealing with the private sector more as contractors than partners.

There are signs that the transformation has started, but much more needs to be done to encourage and to enable the private sector to play a more active part in the development of space commercialization.

12.   Postscript

It is barely thirty years since the first, primitive satellite orbited the Earth, and already many of the developments of space technology are taken for granted. The weather charts which appear daily on television have long ceased to cause viewers to marvel at the wonders of the satellite. Successful launches of even the most sophisticated rockets have little chance of appearing in the news, other than as a two-liner; an accident or even a serious launch delay, of course, still merits a few columns and even a photograph or two.

Perhaps one should not complain at this indifference. Space has made its own way quite creditably over the past three decades, and one can be forgiven for believing that it has no need for publicity except in journals produced by and for the space community. Certainly the telecommunications sector is so largely privatized that it must be treated like any other industry struggling to make a profit.

The future development of space, however, will not depend crucially on the investment of satcom companies; it still needs major contributions from government funds. For this reason many governments are asking, some not for the first time, whether the investment is justified. For many years it has sufficed to reply that space is the last frontier or the tallest mountain waiting for human conquest. The emotive analogies have not been lacking. Increasingly, however, some governments have been seeking a more substantial and even a quantified answer to their question: 'What are we getting for our money?'. It is not easy to give a satisfactory reply—at least not one which might be regarded as satisfactory by a Ministry of Finance official.

The difficulties start when one examines the individual space sectors one by one. In each case it is possible to marshal arguments showing that the investment would produce the same—or greater—benefits if it was directed to the corresponding ground activity. Thus the ground astronomer would generally argue in favour of putting most of the 'space science' money into ground observatories. Even in the field of telecommunications, many experts doggedly prefer underwater cable to satellite connections. Critics of microgravity attack the creation of expensive space-borne laboratories, claiming that better and cheaper results can be obtained on the ground. And so on.

There is, of course, some truth in these anti-space arguments. The main defence must be that space techniques must be made to complement what can be done on the ground, and to replace them only when economics are in favour of a space solution. A great deal of healthy progress has already been made in this direction, and more will be required to defend the public funding which is still needed to maintain progress.

The fact is that, although the various sectors of space activity appear to be independent, they are all more or less dependent on the existence of a solid technological base. This base, which inevitably includes the means of launching spacecraft and all the costly ground equipment needed for space programmes, cannot yet be financed entirely from private sector use of space. The individual sectors, even telecommunications, depend to a greater or lesser extent on the investments made by the major space agencies. Without this public funding the rate of development would be significantly slower and in some areas would stop altogether.

But would it matter? Yes, it most certainly would. Without going to the extremes of 'the final frontier' variety, there are undoubted benefits to be claimed for space and potentially many more to come. To have a complete perception of the national value of investment in space one must combine elements from across all the sectors—science, meteorology, remote sensing, communications, etc.—and then add those factors which are so difficult to quantify; for example: working at the limits of our technological capability in work which requires the highest standards of quality control and stimulating young people to take an interest in science, advanced technology, and exploration. There are regular reports which indicate the seed effect of space agency money injected into industry; most analysts agree that it produces several times its own value in new business which firms would not have acquired without the added capability resulting from their space work. Less mathematical are the arguments that an attractive space programme is a fine flagship for a trading nation, and that it constitutes a worthwhile aid to foreign policy (particularly in days when Earth observation and surveillance are so important).

The most successful space nations (France and Japan are excellent examples) are those which understand the spectrum of advantages and the interlocking nature of the different aspects of space activity. They spend time working out a detailed plan and they provide government funds to achieve it. The 'stop–go' syndrome is the death of any space programme.

In most countries of the world, new priorities have arisen in the past few years. The East–West *détente*, the removal of the Iron Curtain, and the political acceptance that environmental problems cannot be left to the 'Greens' have each had significant effects on national budgeting. Space still has a major role, or more accurately, several different roles, to play in these new scenarios, but it can no longer be considered an independent, free-standing activity to be left entirely to space agencies. More and more we will see other influential players wishing to have a say in how the space money is spent.

But even at this time, when public space expenditure is under particularly sharp scrutiny, there does not seem to be any danger of the overall size of this investment being reduced in the coming decade. What is highly likely, however, is that the money will be directed away from the larger hardware programmes and towards the use of space techniques in, for example, the protection and improvement of the environment. The proportion of money spent on manned flight, much to the fury of the addicts, is likely to be reduced in the 1990s and reserved for those space operations where human presence is really crucial for success.

It is to be hoped that on both sides of the Atlantic some government money will be reserved for a continuation of the exciting planetary explorations. There is no need to pretend that the final secret of the origin of the universe is just around the corner, and that one more mission will put it in our hands. There is adequate justification for investing an infinitesimal part of our common wealth to a systematic increase in our knowledge commensurate with our newly-won technical prowess.

Common sense—and where this fails, Treasury officials—should dictate that our space programmes maintain a healthy balance between the immediately useful and the exciting. This probably implies that space agencies can no longer be given a free hand in proposing and deciding their new projects; a much more structured consultation with potential end-users is essential. This consultation must start before the space project is crystallized in the minds of the engineers, so that the end product meets the needs of users and at a price they are prepared to pay.

This perhaps paints a duller picture of future space programmes than in the 1960s and 1970s, but space has come of age and, as we all know, adulthood, brings with it responsibilities which do not burden us during our youth. As consolation it should be stressed that the next two or three decades will, in spite of the newly acquired maturity, certainly produce its share of new and stimulating space programmes: not only some form of international space station—exciting enough—but also the emergence of the third generation of launcher—the single stage to orbit, and back. And there will be major scientific breakthroughs: landing on Mars, bringing back samples from comets. There is no need, therefore, for despair: space has by no means lost its power to charm and thrill us, it is simply, so to say, changing into long trousers. Enthusiasts should not be disappointed that my crystal ball has not revealed space colonies, holidays in space ('Round the Moon and back in time for tea') and the like. Yes, these are all on the way, but it is a way which leads through a decade or more of self-disciplined progress designed to put space development on a firmer basis. We still need visionaries, but a few good system engineers and cost accountants will be necessary to make the progress so many of us would like to see.

Index